Praise for

A UNIVERSE FROM NOTHING

"Krauss possesses a rare talent for making the hardest ideas in astrophysics accessible to the layman, due in part to his sly humor . . . one has to hope that this book won't appeal only to the partisans of the culture wars—it's just too good and interesting for that. Krauss is genuinely in awe of the 'wondrously strange' nature of our physical world, and his enthusiasm is infectious."

—**Associated Press**

"An eloquent guide to our expanding universe . . . There have been a number of fine cosmology books published recently but few have gone so far, and none so eloquently, in exploring why it is unnecessary to invoke God to light the blue touchpaper and set the universe in motion."

—*Financial Times*

"How physicists came up with the current model of the cosmos is quite a story, and to tell it in his elegant *A Universe from Nothing*, physicist Lawrence Krauss walks a carefully laid path . . . It would be easy for this remarkable story to revel in self-congratulation, but Krauss steers it soberly and with grace . . . His asides on how he views each piece of science and its chances of being right are refreshingly honest . . . unstable nothingness, as described by Krauss . . . is also invigorating for the rest of us, because in this nothingness there are many wonderful things to see and understand."

—*Nature*

"[An] excellent guide to cutting-edge physics . . . As Krauss elegantly argues in *A Universe from Nothing*, the accelerating expansion, indeed the whole existence of the cosmos, is most likely powered by 'nothing.' Krauss is an exemplary interpreter of tough science, and the central part of the book, where he discusses what we know about the history of the universe—and how we know it—is perfectly judged. It is detailed but lucid, thorough but not stodgy . . . Space and time can indeed come from nothing; nothing, as Krauss explains beautifully, being an extremely unstable state from which the production of "'something'" is pretty much inevitable . . . *A Universe from Nothing* is a great book: readable, informative and topical."

—*New Scientist*

"With its mind-bending mechanics, Krauss argues, our universe may indeed have appeared from nowhere, rather than at the hands of a divine creator. There's some intellectual heavy lifting here—Einstein is the main character, after all—but the concepts are articulated clearly, and the thrill of discovery is contagious. 'We are like the early terrestrial mapmakers,' Krauss writes, puzzling out what was once solely the province of our imaginations."

—*Mother Jones*

"His arguments for the birth of the universe out of nothingness from a physical, rather than theological, beginning not only are logical but celebrate the wonder of our natural universe. Recommended."

—*Library Journal*

"Lively and humorous as well as informative . . . Readers will find the result of Krauss's '[celebration of our] absolutely surprising and fascinating universe' as compelling as it is intriguing."

—*Publishers Weekly*

"The author delivers plenty of jolts in this enthusiastic and lucid but demanding overview of the universe, which includes plenty of mysteries—but its origin isn't among them. A thoughtful, challenging book—but not for the faint of heart or those not willing to read carefully."

—*Kirkus Reviews*

"Krauss is a lucid . . . writer, as well as a sparkling speaker and wit, an all-purpose science communicator . . . [I]t is an account of how to untie a paradox, scientifically. And it's also a scientist's hymn—a song of secular appreciation—to the unseen."

—cbcnews.ca

"In *A Universe from Nothing,* Lawrence Krauss has written a thrilling introduction to the current state of cosmology—the branch of science that tells us about the deep past and deeper future of everything. As it turns out, everything has a lot to do with nothing—and nothing to do with God. This is a brilliant and disarming book."

—Sam Harris, author of *The Moral Landscape*

"People always say you can't get something from nothing. Thankfully, Lawrence Krauss didn't listen. In fact, something big happens to you during this book about cosmic nothing, and before you can help it, your mind will be expanding as rapidly as the early universe."

—Sam Kean, author of *The Disappearing Spoon*

"Nothing is not nothing. Nothing is something. That's how a cosmos can be spawned from the void—a profound idea conveyed in *A Universe From Nothing* that unsettles some yet enlightens others. Meanwhile, it's just another day on the job for physicist Lawrence Krauss."

—Neil deGrasse Tyson, astrophysicist, American Museum of Natural History

"With characteristic wit, eloquence, and clarity Lawrence Krauss gives a wonderfully illuminating account of how science deals with one of the biggest questions of all: how the universe's existence could arise from nothing. It is a question that philosophy and theology get themselves into muddle over, but that science can offer real answers to, as Krauss's lucid explanation shows. Here is the triumph of physics over metaphysics, reason and enquiry over obfuscation and myth, made plain for all to see: Krauss gives us a treat as well as an education in fascinating style."

—A. C. Grayling, author of *The Good Book*

"Astronomers at the beginning of the twentieth century were wondering whether there was anything beyond our Milky Way Galaxy. As Lawrence Krauss lucidly explains, astronomers living two trillion years from now, will perhaps be pondering precisely the same question! Beautifully navigating through deep intellectual waters, Krauss presents the most recent ideas on the nature of our cosmos, and of our place within it. A fascinating read."

—Mario Livio, author of *Is God A Mathematician?* and *The Golden Ratio*

"In this clear and crisply written book, Lawrence Krauss outlines the compelling evidence that our complex cosmos has evolved from a hot, dense state and how this progress has emboldened theorists to develop fascinating speculations about how things really began."

—Sir Martin Rees, author of *Our Final Hour*

"A series of brilliant insights and astonishing discoveries have rocked the Universe in recent years, and Lawrence Krauss has been in the thick of it. With his characteristic verve, and using many clever devices, he's made that remarkable story remarkably accessible. The climax is a bold scientific answer to the great question of existence: Why is there something rather than nothing?"

—Frank Wilczek, Nobel Laureate and Herman
Feshbach professor of Physics at MIT, and
author of *The Lightness of Being*

Also by Lawrence M. Krauss

The Fifth Essence

Fear of Physics

The Physics of Star Trek

Beyond Star Trek:
From Alien Invasions to the End of Time

Quintessence:
The Mystery of the Missing Mass

Atom:
A Single Oxygen Atom's Journey from the Big Bang
to Life on Earth . . . and Beyond

Hiding in the Mirror:
The Quest for Alternate Realities, from Plato to String Theory

Quantum Man:
Richard Feynman's Life in Science

A UNIVERSE FROM NOTHING

*Why There Is Something
Rather than Nothing*

LAWRENCE M. KRAUSS

With an Afterword by **Richard Dawkins**

ATRIA PAPERBACK

New York   London   Toronto   Sydney   New Delhi

ATRIA PAPERBACK
A Division of Simon & Schuster, Inc.
1230 Avenue of the Americas
New York, NY 10020

First Atria Books trade paperback edition March 2013

ATRIA PAPERBACK and colophon are trademarks of Simon & Schuster, Inc.

For information about special discounts for bulk purchases,
please contact Simon & Schuster Special Sales at 1-866-506-1949
or business@simonandschuster.com.

The Simon & Schuster Speakers Bureau can bring authors to your live event.
For more information or to book an event contact the Simon & Schuster Speakers
Bureau at 1-866-248-3049 or visit our website at www.simonspeakers.com.

Manufactured in the United States of America

20 19 18 17

The Library of Congress has cataloged the hardcover edition as follows:
Krauss, Lawrence Maxwell.
A universe from nothing : why there is something rather than nothing /
Lawrence M. Krauss ; with an afterword by Richard Dawkins.
p.   cm.
Includes index.
1. Cosmology. 2. Beginning. 3. End of the universe. I. Title.
QB981.K773 2012   2011032519
523.1'8—dc23

ISBN 978-1-4516-2445-8
ISBN 978-1-4516-2446-5 (pbk)
ISBN 978-1-4516-2447-2 (ebook)

To Thomas, Patty, Nancy, and Robin,
for helping inspire me to create something
from nothing . . .

On this site in 1897,
Nothing happened.
—Plaque on wall of Woody Creek Tavern,
Woody Creek, Colorado

CONTENTS

Preface to the
Paperback Edition

Since the hardcover version of this book first appeared, a visceral negative reaction among some commentators to the very idea of a universe arising from nothing has been balanced by a major scientific discovery that supports this possibility. The confirmation of the Higgs boson refines our understanding of the relationship between seemingly empty space and our existence. I want to elaborate on both the Higgs boson and the negative reactions to *A Universe from Nothing* in this new preface.

When I chose to subtitle this book *Why There Is Something Rather Than Nothing*, I wanted to connect the remarkable discoveries of modern science to a question that has fascinated theologians, philosophers, natural philosophers, and the general public for more than two millennia. But I wasn't fully aware of how my choice of words might lead to the same kind of confusion that occurs whenever one says in public that Evolution is a *theory*.

In popular parlance, *theory* means something very different from its scientific sense. So too *nothing* is a hot-button issue for some people, a line in the sand that some people are not willing to cross, so that even using the word, just as using the word *God*, can be so polarizing that it obfuscates more important issues. A similar remark can be made about the question

"Why?": using *why* and *nothing* together can be as explosive as mixing diesel fuel and fertilizer.

In chapter 9 of this book I mention a fact that I now want to introduce first here. Whenever one asks "Why?" in science, one actually means "How?". "Why?" is not really a sensible question in science because it usually implies purpose and, as anyone who has been the parent of a small child knows, one can keep on asking "Why?" forever, no matter what the answer to the previous question. Ultimately, the only way to end the conversation seems to be to say "Because!"

Science *changes* the meaning of questions, especially why-like questions, as it progresses. Here is an early example of this fact, which illustrates a number of features in common with the more recent revelations I treat in this book.

The renowned astronomer Johannes Kepler claimed in 1595 to have had an epiphany when he suddenly thought he had answered a profoundly important why question: "Why are there six planets?" The answer, he believed, lay in the view of the five Platonic solids, those sacred objects from geometry whose faces can be composed of regular polygons—triangles, squares, etc.—and that could be circumscribed by spheres whose size would increase as the number of faces of the solid increased. If these spheres then separated the orbits of the six known planets, he conjectured, perhaps their relative distances from the sun and the fact that there were just six of them could be understood as revealing, in a profound and deep sense, the mind of God, the mathematician. (The idea that geometry was sacred goes back as far as Pythagoras.) "Why are there six planets?"—then, in 1595—was considered a meaningful question, one that revealed purpose to the universe.

Now, however, we understand the question is meaningless. In the first place, we know there are not six planets, there are nine planets. (Pluto will always be a planet for me. Not only do I like to annoy my friend Neil deGrasse Tyson by so insisting, but my daughter did her fourth-grade science project on Pluto, and I don't want that to have been in vain!) More important, however, we know our solar system is *not* unique, which Kepler and his era did not know. More than two thousand planets orbiting other stars have been discovered (by a satellite named Kepler, coincidentally!).

The important question then becomes not "*Why?*" but "*How* does our solar system have nine planets?" (or, eight planets, depending upon your count). Since clearly lots of different solar systems exist, with very different features, what we really want to know is how typical we are, what specific conditions might have existed allowing our solar system to have four rocky planets closest to the sun, surrounded by a number of far larger gas giants. The answer to this question might shed light on the likelihood of finding life elsewhere in the universe, for example.

Most important, however, we realize that there is nothing profound about six (or eight or nine), nothing that points to purpose or design . . . no evidence of "purpose" in the distribution of planets in the universe. Not only has "why" become "how" but "why" no longer has any verifiable meaning.

So too, when we ask "Why is there something rather than nothing?" we really mean "How is there something rather than nothing?" This brings me to the second confusion engendered by my choice of words. There are many seeming "miracles" of nature that appear so daunting that many have given up trying to find an explanation of how we came to be and, instead, blame it all on God. But the question I really care about, and the one that science can actually address, is the question of how all the "stuff" in the universe could have come from no "stuff," and how, if you wish, formlessness led to form. That is what seems so astounding and nonintuitive. It seems to violate everything we know about the world—in particular the fact that energy in its various forms, including mass, is conserved. Common sense suggests that "nothing," in this sense the absence of "something," should have zero total energy. Therefore, where did the 400 billion or so galaxies that make up the observable universe come from?

The fact that we need to refine what we mean by "common sense" in order to accommodate our understanding of nature is, to me, one of the most remarkable and liberating aspects of science. Reality liberates us from the biases and misconceptions that have arisen because our intellects evolved through our animal ancestors, whose survival was based on whether predators might lurk behind trees or in caves and not on understanding the wave function of electrons in atoms.

Our modern conception of the universe is so foreign to what even scientists generally believed a mere century ago that it is a tribute to the power of the scientific method and the creativity and persistence of humans who want to understand it. That is worth celebrating. As I describe in this book, the question and the possible answers to how something might come from nothing are even more interesting than merely the possibility of galaxies manifesting from empty space. Science provides a possible road map for the creation of space (and time) itself—and perhaps also an understanding of how the laws of physics that govern the dynamics of space and time can arise haphazardly.

For many people, however, the fascinating possible resolutions of these age-old mysteries are not sufficient. The deeper question of nonexistence overwhelms them. Can we understand how absolute nothingness, without even the potential for anything at all to exist, does not still reign supreme? Can one ever say anything other than the fact that the nothing that became our something was a part of "something" else, in which the potential for our existence, or any existence, was always implicit?

In the book I take a rather flippant attitude toward this concern, because I don't think it adds anything to the productive discussion, which is "What questions are actually answerable by probing the universe?" I have discounted this philosophical issue, but not because I think those people who occupy themselves with certain aspects of it are not trying hard to define logical questions. Rather, I discount this aspect of philosophy here because I think it bypasses the really interesting and answerable physical questions associated with the origin and evolution of our universe. No doubt some will view this as my own limitation, and maybe it is. But it is within that context that people should read this book. I don't make any claims to answer any questions that science cannot answer, and I have tried very carefully within the text to define what I mean by "nothing" and "something." If those definitions differ from those you would like to adopt, so be it. Write your own book. But don't discount the remarkable human adventure that is modern science because it doesn't console you.

Now, the good news! This past summer, physicists around the world, including me, were glued to computers at very odd hours to watch live as scientists at the Large Hadron Collider, outside Geneva, announced that they had found one of the most important missing pieces of the jigsaw puzzle that is nature—the Higgs particle (or Higgs boson).

Proposed almost fifty years ago to allow for consistency between theoretical predictions and experimental observations in elementary particle physics, the Higgs particle's discovery caps one of the most remarkable intellectual adventures in human history—one that anyone interested in the progress of knowledge should at least be aware of—and makes even more remarkable the precarious accident that allowed our existence to form from nothing, the subject of this book. The discovery is further proof that the universe of our senses is just the tip of a vast, largely hidden cosmic iceberg and that seemingly empty space can provide the seeds for our existence.

The prediction of the Higgs particle accompanied a remarkable revolution that completely changed our understanding of particle physics in the latter part of the twentieth century. Just fifty years ago, in spite of the great advances of physics in the previous half century, we understood only one of the four fundamental forces of nature—electromagnetism—as a fully consistent quantum theory. In just one subsequent decade, however, not only had three of the four known forces surrendered to our investigations, but a new elegant unity of nature had been uncovered. It was found that all of the known forces could be described using a single mathematical framework—and that two of the forces, electromagnetism and the weak force (which governs the nuclear reactions that power the sun), were actually different manifestations of a single underlying force.

How could two such different forces be related? After all, the photon, the particle that conveys electromagnetism, has no mass, while the particles that convey the weak force are very massive—almost one hundred times as heavy as the particles that make up atomic nuclei, a fact that explains why the weak force is weak.

British physicist Peter Higgs and several others showed that, if there exists an otherwise invisible background field (Higgs field) permeating all of space, then the particles that convey some force like electromagnetism can interact with this field and effectively encounter resistance to their motion and slow down, like a swimmer moving through molasses. As a result, these particles can behave as if they are heavy, as if they have a mass. The physicist Steven Weinberg (and somewhat later Abdus Salam) applied this idea to a model of the weak and electromagnetic forces previously proposed by Sheldon L. Glashow, and everything fit together.

This idea can be extended to the rest of particles in nature, including particles like those that make up the protons and neutrons, as well as fundamental particles like electrons, all of which combine to make up the atoms in our bodies. If some particle interacts more strongly with this background field, it ends up acting heavier. If it interacts more weakly, it acts lighter. If it doesn't interact at all, like the photon, it remains massless.

If anything sounds too good to be true, this is it. The miracle of mass—indeed, of our very existence (because if not for the Higgs, there would be no stars, no planets, and no people)— is apparently possible because of some otherwise hidden background field whose only effect seems to be to allow the world to look the way it does.

But relying on invisible miracles is the stuff of religion, not science. To ascertain whether this remarkable accident was real, physicists relied on another facet of the quantum world. Associated with every background field is a particle, and if you pick a point in space and hit it hard enough, you may whack out real particles. The trick is hitting it hard enough over a small enough volume. And that's the rub. After fifty years of trying, including a failed attempt in the United States to build an accelerator to test these ideas, no sign of the Higgs had appeared. In fact, I was betting against it, since a career in theoretical physics has taught me that nature usually has a far richer imagination than we do.

Until July.

The apparent discovery of the Higgs boson may not result in a better toaster or a faster car. But it provides a remarkable celebration of the human mind's capacity to uncover nature's secrets, and of the technology we have built to control them. Hidden in what seems like empty space—indeed, like nothing— appear to be the very elements that allow for our existence.

The discovery of a Higgs field further validates many of the ideas I discuss in this book. The idea that the very early universe went through a period of superluminal expansion, called inflation, that basically produced almost all the space and matter in the observable universe from almost nothing relies heavily on the possibility that another field, much like the Higgs field we seem to have discovered this past year, momentarily held sway in early times.

The existence of a Higgs field permeating all of space today also begs several important questions, most notably "What conditions in the early universe led to such a cosmic accident?" "Why does the field have the value it is measured to have?" "Could it have been different?" "Could the laws of physics, had initial conditions been slightly different, have resulted in a universe without matter as we observe it today?" These are precisely the kind of questions I discuss near the end of this book.

Whatever the ultimate resolution of these puzzles, and others that I shall discuss in this book, our discoveries in fundamental physics and astronomy over the past forty years have changed our understanding of our place in the universe in profound ways, by changing not only the questions we ask, but the very meaning of the questions we have asked. That, as I want to stress once again, is perhaps the greatest legacy of modern science, a legacy it shares with great music, great literature, and great art, and one that needs to be shared more widely.

PREFACE

Dream or nightmare, we have to live our experience as it is, and we have to live it awake. We live in a world which is penetrated through and through by science and which is both whole and real. We cannot turn it into a game simply by taking sides.

—Jacob Bronowski

In the interests of full disclosure right at the outset I must admit that I am not sympathetic to the conviction that creation requires a creator, which is at the basis of all of the world's religions. Every day beautiful and miraculous objects suddenly appear, from snowflakes on a cold winter morning to vibrant rainbows after a late-afternoon summer shower. Yet no one but the most ardent fundamentalists would suggest that each and every such object is lovingly and painstakingly and, most important, purposefully created by a divine intelligence. In fact, many laypeople as well as scientists revel in our ability to explain how snowflakes and rainbows can spontaneously appear, based on simple, elegant laws of physics.

Of course, one can ask, and many do, "Where do the laws of physics come from?" as well as more suggestively, "Who created these laws?" Even if one can answer this first query, the petitioner will then often ask, "But where did that come from?" or "Who created that?" and so on.

Ultimately, many thoughtful people are driven to the apparent need for First Cause, as Plato, Aquinas, or the modern Roman Catholic Church might put it, and thereby to suppose some divine being: a creator of all that there is, and all that there ever will be, someone or something eternal and everywhere.

Nevertheless, the declaration of a First Cause still leaves open the question, "Who created the creator?" After all, what is the difference between arguing in favor of an eternally existing creator versus an eternally existing universe without one?

These arguments always remind me of the famous story of an expert giving a lecture on the origins of the universe (sometimes identified as Bertrand Russell and sometimes William James), who is challenged by a woman who believes that the world is held up by a gigantic turtle, who is then held up by another turtle, and then another . . . with further turtles "all the way down!" An infinite regress of some creative force that begets itself, even some imagined force that is greater than turtles, doesn't get us any closer to what it is that gives rise to the universe. Nonetheless, this metaphor of an infinite regression may actually be closer to the real process by which the universe came to be than a single creator would explain.

Defining away the question by arguing that the buck stops with God may seem to obviate the issue of infinite regression, but here I invoke my mantra: The universe is the way it is, whether we like it or not. The existence or nonexistence of a creator is independent of our desires. A world without God or purpose may seem harsh or pointless, but that alone doesn't require God to actually exist.

Similarly, our minds may not be able to easily comprehend infinities (although mathematics, a product of our minds, deals with them rather nicely), but that doesn't tell us that infinities don't exist. Our universe could be infinite in spatial or temporal extent. Or, as Richard Feynman once put it, the laws of physics could be like an infinitely layered onion, with new laws becoming operational as we probe new scales. *We simply don't know!*

For more than two thousand years, the question, "Why is there something rather than nothing?" has been presented as a challenge to the proposition that our universe—which contains the vast complex of stars, galaxies, humans, and who knows what else—might have arisen without design, intent, or purpose. While this is usually framed as a philosophical or religious question, it is first and foremost a question about the natural world, and so the appropriate place to try and resolve it, first and foremost, is with science.

The purpose of this book is simple. I want to show how modern science, in various guises, can address and *is* addressing the question of why there is something rather than nothing: The answers that have been obtained—from staggeringly beautiful experimental observations, as well as from the theories that underlie much of modern physics—all suggest that getting something from nothing is not a problem. Indeed, something from nothing may have been *required* for the universe to come into being. Moreover, all signs suggest that this is how our universe *could* have arisen.

I stress the word *could* here, because we may never have enough empirical information to resolve this question unambiguously. But the fact that a universe from nothing is even plausible is certainly significant, at least to me.

Before going further, I want to devote a few words to the notion of "nothing"—a topic that I will return to at some length later. For I have learned that, when discussing this question in public forums, nothing upsets the philosophers and theologians who disagree with me more than the notion that I, as a scientist, do not truly understand "nothing." (I am tempted to retort here that theologians are experts at nothing.)

"Nothing," they insist, is not any of the things I discuss. Nothing is "nonbeing," in some vague and ill-defined sense. This reminds me of my own efforts to define "intelligent design" when I first began debating with creationists, of which, it became clear, there is no clear definition, except to say what it isn't. "Intelligent design" is simply a unifying umbrella for opposing evolution. Similarly, some philosophers and many theologians define and redefine "nothing" as not being any of the versions of nothing that scientists currently describe.

But therein, in my opinion, lies the intellectual bankruptcy of much of theology and some of modern philosophy. For surely "nothing" is every bit as physical as "something," especially if it is to be defined as the "absence of something." It then behooves us to understand precisely the physical nature of both these quantities. And without science, any definition is just words.

A century ago, had one described "nothing" as referring to purely empty space, possessing no real material entity, this might have received little argument. But the results of the past century have taught us that empty space is in fact far from the inviolate nothingness that we presupposed before we learned more about how nature works. Now, I am told by religious critics that I cannot refer to empty space as "nothing," but rather as a "quantum vacuum," to distinguish it from the philosopher's or theologian's idealized "nothing."

So be it. But what if we are then willing to describe "nothing" as the absence of space and time itself? Is this sufficient? Again, I suspect it would have been . . . at one time. But, as I shall describe, we have learned that space and time can themselves spontaneously appear, so now we are told that even this "nothing" is not really the nothing that matters. And we're told that the escape from the "real" nothing requires divinity, with "nothing" thus defined by fiat to be "that from which only God can create something."

It has also been suggested by various individuals with whom I have debated the issue that, if there is the "potential" to create something, then that is not a state of true nothingness. And surely having laws of nature that give such potential takes us away from the true realm of nonbeing. But then, if I argue that perhaps the laws themselves also arose spontaneously, as I shall describe might be the case, then that too is not good enough, because whatever system in which the laws may have arisen is not true nothingness.

Turtles all the way down? I don't believe so. But the turtles are appealing because science is changing the playing field in ways that make people uncomfortable. Of course, that is one of the purposes of science (one might have said "natural philosophy" in Socratic times). Lack of comfort means we are on the threshold of new insights. Surely, invoking "God" to avoid difficult questions of "how" is merely intellectually lazy. After all, if there were no potential for creation, then God couldn't have created anything. It would be semantic hocus-pocus to assert that the potentially infinite regression is avoided because God exists outside nature and, therefore, the "potential" for existence itself is not a part of the nothingness from which existence arose.

My real purpose here is to demonstrate that in fact science *has* changed the playing field, so that these abstract and useless debates about the nature of nothingness have been replaced by useful, operational efforts to describe how our universe might actually have originated. I will also explain the possible implications of this for our present and future.

This reflects a very important fact. When it comes to understanding how our universe evolves, religion and theology have been at best irrelevant. They often muddy the waters, for example, by focusing on questions of nothingness without providing any definition of the term based on empirical evidence. While we do not yet fully understand the origin of our universe, there is no reason to expect things to change in this regard. Moreover, I expect that ultimately the same will be true for our understanding of areas that religion now considers its own territory, such as human morality.

Science has been effective at furthering our understanding of nature because the scientific ethos is based on three key principles: (1) follow the evidence wherever it leads; (2) if one has a theory, one needs to be willing to try to prove it wrong as much as one tries to prove that it is right; (3) the ultimate arbiter of truth is experiment, not the comfort one derives from one's a priori beliefs, nor the beauty or elegance one ascribes to one's theoretical models.

The results of experiments that I will describe here are not only timely, they are also unexpected. The tapestry that science weaves in describing the evolution of our universe is far richer and far more fascinating than any revelatory images or imaginative stories that humans have concocted. Nature comes up with surprises that far exceed those that the human imagination can generate.

Over the past two decades, an exciting series of developments in cosmology, particle theory, and gravitation have completely changed the way we view the universe, with startling and profound implications for our understanding of its origins as well as its future. Nothing could therefore not be more interesting to write about, if you can forgive the pun.

The true inspiration for this book comes not so much from

a desire to dispel myths or attack beliefs, as from my desire to celebrate knowledge and, along with it, the absolutely surprising and fascinating universe that ours has turned out to be.

Our search will take us on a whirlwind tour to the farthest reaches of our expanding universe, from the earliest moments of the Big Bang to the far future, and will include perhaps the most surprising discovery in physics in the past century.

Indeed, the immediate motivation for writing this book now is a profound discovery about the universe that has driven my own scientific research for most of the past three decades and that has resulted in the startling conclusion that most of the energy in the universe resides in some mysterious, now inexplicable form permeating all of empty space. It is not an understatement to say that this discovery has changed the playing field of modern cosmology.

For one thing, this discovery has produced remarkable new support for the idea that our universe arose from precisely nothing. It has also provoked us to rethink both a host of assumptions about the processes that might govern its evolution and, ultimately, the question of whether the very laws of nature are truly fundamental. Each of these, in its own turn, now tends to make the question of why there is something rather than nothing appear less imposing, if not completely facile, as I hope to describe.

The direct genesis of this book hearkens back to October of 2009, when I delivered a lecture in Los Angeles with the same title. Much to my surprise, the YouTube video of the lecture, made available by the Richard Dawkins Foundation, has since become something of a sensation, with more than 1.5 million viewings as of this writing, and numerous copies of parts of it being used by both the atheist and theist communities in their debates.

Because of the clear interest in this subject, and also as a result of some of the confusing commentary on the web and in various media following my lecture, I thought it worth producing a more complete rendition of the ideas that I had expressed there in this book. Here I can also take the opportunity to add to the arguments I presented at the time, which focused almost completely on the recent revolutions in cosmology that have changed our picture of the universe, associated with the discovery of the energy and geometry of space, and which I discuss in the first two-thirds of this book.

In the intervening period, I have thought a lot more about the many antecedents and ideas constituting my argument; I've discussed it with others who reacted with a kind of enthusiasm that was infectious; and I've explored in more depth the impact of developments in particle physics, in particular, on the issue of the origin and nature of our universe. And finally, I have exposed some of my arguments to those who vehemently oppose them, and in so doing have gained some insights that have helped me develop my arguments further.

While fleshing out the ideas I have ultimately tried to describe here, I benefitted tremendously from discussions with some of my most thoughtful physics colleagues. In particular I wanted to thank Alan Guth and Frank Wilczek for taking the time to have extended discussions and correspondence with me, resolving some confusions in my own mind and in certain cases helping reinforce my own interpretations.

Emboldened by the interest of Leslie Meredith and Dominick Anfuso at Simon & Schuster in the possibility of a book on this subject, I then contacted my friend Christopher Hitchens, who, besides being one of the most literate and brilliant individuals I know, had himself been able to use some of the arguments from my lecture in his remarkable series of debates on science and religion. Christopher, in spite of his ill health, kindly, generously, and bravely agreed to write a foreword. For that act of friendship and trust, I will be eternally grateful. Unfortunately, Christopher's illness eventually overwhelmed him to the extent that completing the foreword became impossible, in spite of his best efforts, and he tragically passed away just before the first edition of this book appeared. I miss him and the world is emptier without him. Nevertheless, in an embarrassment of riches, my eloquent, brilliant friend, the renowned scientist and writer Richard Dawkins, had earlier agreed to write an afterword. After my first draft was completed, he then proceeded to produce something in short order whose beauty and clarity was astounding, and at the same time humbling. I remain in awe. To Christopher, Richard, then, and all of those above, I issue my thanks for their support and encouragement, and for motivating me to once again return to my computer and write.

CHAPTER 1

A COSMIC MYSTERY STORY:
BEGINNINGS

The Initial Mystery that attends any journey is: how did the traveler reach his starting point in the first place?
—LOUISE BOGAN, *Journey Around My Room*

It was a dark and stormy night.

Early in 1916, Albert Einstein had just completed his greatest life's work, a decade-long, intense intellectual struggle to derive a new theory of gravity, which he called the general theory of relativity. This was not just a new theory of gravity, however; it was a new theory of space and time as well. And it was the first scientific theory that could explain not merely how objects move through the universe, but also how the universe itself might evolve.

There was just one hitch, however. When Einstein began to apply his theory to describing the universe as a whole, it became clear that the theory didn't describe the universe in which we apparently lived.

Now, almost one hundred years later, it is difficult to fully appreciate how much our picture of the universe has changed in the span of a single human lifetime. As far as the scientific community in 1917 was concerned, the universe was static and eternal, and consisted of a single galaxy, our Milky Way, surrounded by a vast, infinite, dark, and empty space. This is, after all, what you would guess by looking up at the night sky with your eyes, or with a small telescope, and at the time there was little reason to suspect otherwise.

In Einstein's theory, as in Newton's theory of gravity before it, gravity is a purely attractive force between all objects. This means that it is impossible to have a set of masses located in space at rest forever. Their mutual gravitational attraction will ultimately cause them to collapse inward, in manifest disagreement with an apparently static universe.

The fact that Einstein's general relativity didn't appear consistent with the then picture of the universe was a bigger blow to him than you might imagine, for reasons that allow me to dispense with a myth about Einstein and general relativity that has always bothered me. It is commonly assumed that Einstein worked in isolation in a closed room for years, using pure thought and reason, and came up with his beautiful theory, independent of reality (perhaps like some string theorists nowadays!). However, nothing could be further from the truth.

Einstein was always guided deeply by experiments and observations. While he performed many "thought experiments" in his mind and did toil for over a decade, he learned new mathematics and followed many false theoretical leads in the process before he ultimately produced a theory that was indeed mathematically beautiful. The single most important moment in establishing his love affair with general relativity, however, had to do with observation. During the final hectic weeks that he was completing his theory, competing with the German mathematician David Hilbert, he used his equations to calculate the prediction for what otherwise might seem an obscure astrophysical result: a slight precession in the "perihelion" (the point of closest approach) of Mercury's orbit around the Sun.

Astronomers had long noted that the orbit of Mercury departed slightly from that predicted by Newton. Instead of being a perfect ellipse that returned to itself, the orbit of Mercury precessed (which means that the planet does not return precisely to the same point after one orbit, but the orientation of the ellipse shifts slightly each orbit, ultimately tracing out a kind of spiral-like pattern) by an incredibly small amount: 43 arc seconds (about $\frac{1}{100}$ of a degree) per century.

When Einstein performed his calculation of the orbit using his theory of general relativity, the number came out just right. As described by an Einstein biographer, Abraham Pais: "This discovery was, I believe, by far the strongest emotional experience in Einstein's scientific life, perhaps in all his life." He claimed to have heart palpitations, as if "something had snapped" inside. A month later, when he described his theory to a friend as one of "incomparable beauty," his pleasure over the mathematical form was indeed manifest, but no palpitations were reported.

The apparent disagreement between general relativity and observation regarding the possibility of a static universe did not last long, however. (Even though it did cause Einstein to introduce a modification to his theory that he later called his biggest blunder. But more about that later.) Everyone (with the exception of certain school boards in the United States) now knows that the universe is not static but is expanding and that the expansion began in an incredibly hot, dense Big Bang approximately 13.72 billion years ago. Equally important, we know that our galaxy is merely one of perhaps 400 billion galaxies in the observable universe. We are like the early terrestrial mapmakers, just beginning to fully map the universe on its largest scales. Little wonder that recent decades have witnessed revolutionary changes in our picture of the universe.

The discovery that the universe is not static, but rather expanding, has profound philosophical and religious significance, because it suggested that our universe had a beginning. A beginning implies creation, and creation stirs emotions. While it took several decades following the discovery in 1929 of our expanding universe for the notion of a Big Bang to achieve independent empirical confirmation, Pope Pius XII heralded it in 1951 as evidence for Genesis. As he put it:

> It would seem that present-day science, with one sweep back across the centuries, has succeeded in bearing witness to the august instant of the primordial Fiat Lux [Let there be Light], when along with matter, there burst forth from nothing a sea of light and radiation, and the elements split and churned and formed into millions of galaxies. Thus, with that concreteness which is characteristic of physical proofs, [science] has confirmed the contingency of the universe and also the well-founded deduction as to the epoch when the world came forth from the hands of the Creator. Hence, creation took place. We say: "Therefore, there is a Creator. Therefore, God exists!"

The full story is actually a little more interesting. In fact, the first person to propose a Big Bang was a Belgian priest and physicist named Georges Lemaître. Lemaître was a remarkable combination of proficiencies. He started his studies as an engineer, was a decorated artilleryman in World War I, and then switched to mathematics while studying for the priesthood in the early 1920s. He then moved on to cosmology, studying first with the famous British astrophysicist Sir Arthur Stanley Eddington before moving on to Harvard and eventually receiving a second doctorate, in physics from MIT.

In 1927, before receiving his second doctorate, Lemaître had actually solved Einstein's equations for general relativity and demonstrated that the theory predicts a nonstatic universe and in fact suggests that the universe we live in is expanding. The notion seemed so outrageous that Einstein himself colorfully objected with the statement "Your math is correct, but your physics is abominable."

Nevertheless, Lemaître powered onward, and in 1930 he further proposed that our expanding universe actually began as an infinitesimal point, which he called the "Primeval Atom" and that this beginning represented, in an allusion to Genesis perhaps, a "Day with No Yesterday."

Thus, the Big Bang, which Pope Pius so heralded, had first been proposed by a priest. One might have thought that Lemaître would have been thrilled with this papal validation, but he had already dispensed in his own mind with the notion that this scientific theory had theological consequences and had ultimately removed a paragraph in the draft of his 1931 paper on the Big Bang remarking on this issue.

Lemaître in fact later voiced his objection to the pope's 1951 claimed proof of Genesis via the Big Bang (not least because he realized that if his theory was later proved incorrect, then the Roman Catholic claims for Genesis might be contested). By this time, he had been elected to the Vatican's Pontifical Academy, later becoming its president. As he put it, "As far as I can see, such a theory remains entirely outside of any metaphysical or religious question." The pope never again brought up the topic in public.

There is a valuable lesson here. As Lemaître recognized, whether or not the Big Bang really happened is a scientific question, not a theological one. Moreover, even if the Big Bang had happened (which all evidence now overwhelmingly supports), one could choose to interpret it in different ways depending upon one's religious or metaphysical predilections. You can choose to view the Big Bang as suggestive of a creator if you feel the need or instead argue that the mathematics of general relativity explain the evolution of the universe right back to its beginning without the intervention of any deity. But such a metaphysical speculation is independent of the physical validity of the Big Bang itself and is irrelevant to our understanding of it. Of course, as we go beyond the mere existence of an expanding universe to understand the physical principles that may address its origin, science can shed further light on this speculation and, as I shall argue, it does.

In any case, neither Lemaître nor Pope Pius convinced the scientific world that the universe was expanding. Rather, as in all good science, the evidence came from careful observations, in this case done by Edwin Hubble, who continues to give me great faith in humanity, because he started out as a lawyer and then became an astronomer.

Hubble had earlier made a significant breakthrough in 1925 with the new Mount Wilson 100-inch Hooker telescope, then the world's largest. (For comparison, we are now building telescopes more than ten times bigger than this in diameter and one hundred times bigger in area!) Up until that time, with the telescopes then available, astronomers were able to discern fuzzy images of objects that were not simple stars in our galaxy. They called these nebulae, which is basically Latin for "fuzzy thing" (actually "cloud"). They also debated whether these objects were in our galaxy or outside of it.

Since the prevailing view of the universe at the time was that our galaxy was all that there was, most astronomers fell in the "in our galaxy" camp, led by the famous astronomer Harlow

Shapley at Harvard. Shapley had dropped out of school in fifth grade and studied on his own, eventually going to Princeton. He decided to study astronomy by picking the first subject he found in the syllabus to study. In seminal work he demonstrated that the Milky Way was much larger than previously thought and that the Sun was not at its center but simply in a remote, uninteresting corner. He was a formidable force in astronomy and therefore his views on the nature of nebulae held considerable sway.

On New Year's Day 1925, Hubble published the results of his two-year study of so-called spiral nebulae, where he was able to identify a certain type of variable star, called a Cepheid variable star, in these nebulae, including the nebula now known as Andromeda.

First observed in 1784, Cepheid variable stars are stars whose brightness varies over some regular period. In 1908, an unheralded and at the time unappreciated would-be astronomer, Henrietta Swan Leavitt, was employed as a "computer" at the Harvard College Observatory. ("Computers" were women brought in to catalogue the brightness of stars recorded on the observatory's photographic plates; women were not allowed to use the observatory telescopes at the time.) Daughter of a Congregational minister and a descendant of the Pilgrims, Leavitt made an astounding discovery, which she further illuminated in 1912: she noticed that there was a regular relationship between the brightness of Cepheid stars and the period of their variation. Therefore, if one could determine the distance to a single Cepheid of a known period (subsequently determined in 1913), then measuring the brightness of other Cepheids of the same period would allow one to determine the distance to these other stars!

Since the observed brightness of stars goes down inversely with the square of the distance to the star (the light spreads out uniformly over a sphere whose area increases as the square of the distance, and thus since the light is spread out over a bigger sphere, the intensity of the light observed at any point decreases inversely with the area of the sphere), determining the distance to faraway stars has always been the major challenge in astronomy. Leavitt's discovery revolutionized the field. (Hubble himself, who was snubbed for the Nobel Prize, often said Leavitt's work deserved the prize, although he was sufficiently self-serving that he might have suggested it only because he would have been a natural contender to share the prize with her for his later work.) Paperwork had actually begun in the Royal Swedish Academy to nominate Leavitt for the Nobel in 1924 when it was learned that she had died of cancer three years earlier. By dint of his force of personality, knack for self-promotion, and skill as an observer, Hubble would become a household name, while Leavitt, alas, is known only to aficionados of the field.

Hubble was able to use his measurement of Cepheids and Leavitt's period-luminosity relation to prove definitively that the Cepheids in Andromeda and several other nebulae were much too distant to be inside the Milky Way. Andromeda was discovered to be another island universe, another spiral galaxy almost identical to our own, and one of the more than 100 billion other galaxies that, we now know, exist in our observable universe. Hubble's result was sufficiently unambiguous that the astronomical community—including Shapley, who, incidentally, by this time had become director of the Harvard College Observatory, where Leavitt had done her groundbreaking work—quickly accepted the fact that the Milky Way is not all there is around us. Suddenly the size of the known universe had expanded in a single leap by a greater amount than it had in centuries! Its character had changed, too, as had almost everything else.

After this dramatic discovery, Hubble could have rested on his laurels, but he was after bigger fish or, in this case, bigger

galaxies. By measuring ever fainter Cepheids in ever more distant galaxies, he was able to map the universe out to ever-larger
scales. When he did, however, he discovered something else that
was even more remarkable: the universe is expanding!

Hubble achieved his result by comparing the distances for the
galaxies he measured with a different set of measurements from
another American astronomer, Vesto Slipher, who had measured
the spectra of light coming from these galaxies. Understanding
the existence and nature of such spectra requires me to take you
back to the very beginning of modern astronomy.

One of the most important discoveries in astronomy was that star
stuff and Earth stuff are largely the same. It all began, as did many
things in modern science, with Isaac Newton. In 1665, Newton,
then a young scientist, allowed a thin beam of sunlight, obtained
by darkening his room except for a small hole he made in his window shutter, through a prism and saw the sunlight disperse into
the familiar colors of the rainbow. He reasoned that white light
from the sun contained all of these colors, and he was correct.

A hundred fifty years later, another scientist examined the dis-
persed light more carefully, discovered dark bands amidst the colors, and reasoned that these were due to the existence of materials
in the outer atmosphere of the sun that were absorbing light of
certain specific colors or wavelengths. These "absorption lines,"
as they became known, could be identified with wavelengths of
light that were measured to be absorbed by known materials on
Earth, including hydrogen, oxygen, iron, sodium, and calcium.

In 1868, another scientist observed two new absorption lines
in the yellow part of the solar spectrum that didn't correspond
to any known element on Earth. He decided this must be due to
some new element, which he called helium. A generation later,
helium was first isolated on Earth.

Looking at the spectrum of radiation coming from other stars is an important scientific tool for understanding their composition, temperature, and evolution. Starting in 1912, Slipher observed the spectra of light coming from various spiral nebulae and found that the spectra were similar to those of nearby stars—except that all of the absorption lines were shifted by the same amount in wavelength.

This phenomenon was by then understood as being due to the familiar "Doppler effect," named after the Austrian physicist Christian Doppler, who explained in 1842 that waves coming at you from a moving source will be stretched if the source is moving away from you and compressed if it is moving toward you. This is a manifestation of a phenomenon we are all familiar with, and by which I am usually reminded of a Sidney Harris cartoon where two cowboys sitting on their horses out in the plains are looking at a distant train, and one says to the other, "I love hearing that lonesome wail of the train whistle as the magnitude of the frequency changes due to the Doppler effect!" Indeed, a train whistle or an ambulance siren sounds higher if the train or ambulance is moving toward you and lower if it is moving away from you.

It turns out that the same phenomenon occurs for light waves as sound waves, although for somewhat different reasons. Light waves from a source moving away from you, either due to its local motion in space or due to the intervening expansion of space, will be stretched, and therefore appear redder than they would otherwise be, since red is the long-wavelength end of the visible spectrum, while waves from a source moving toward you will be compressed and appear bluer.

Slipher observed in 1912 that the absorption lines from the light coming from all the spiral nebulae were almost all shifted systematically toward longer wavelengths (although some, like Andromeda, were shifted toward shorter wavelengths). He correctly inferred that most of these objects therefore were moving away from us with considerable velocities.

Hubble was able to compare his observations of the distance of these spiral galaxies (as they were by now known to be) with Slipher's measurements of the velocities by which they were moving away. In 1929, with the help of a Mount Wilson staff member, Milton Humason (whose technical talent was such that he had secured a job at Mount Wilson without even having a high school diploma), he announced the discovery of a remarkable empirical relationship, now called Hubble's law: There is a linear relationship between recessional velocity and galaxy distance. Namely, galaxies that are ever more distant are moving away from us with faster velocities!

When first presented with this remarkable fact—that almost all galaxies are moving away from us, and those that are twice as far away are moving twice as fast, those that are three times away three times as fast, etc.—it seems obvious what this implies: *We are the center of the universe!*

As some friends suggest, I need to be reminded on a daily basis that *this is not the case*. Rather, it was consistent with precisely the relationship that Lemaître had predicted. Our universe is indeed expanding.

I have tried various ways to explain this, and I frankly don't think there is a good way to do it unless you think outside the box—in this case, outside the universal box. To see what Hubble's law implies, you need to remove yourself from the myopic vantage point of our galaxy and look at our universe from the outside. While it is hard to stand outside a three-dimensional universe, it is easy to stand outside a two-dimensional one. On the next page I have drawn one such expanding universe at two different times. As you can see, the galaxies are farther apart at the second time.

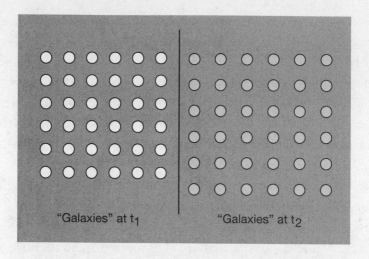

"Galaxies" at t_1 "Galaxies" at t_2

Now imagine that you are living in one of the galaxies at the second time, t_2 which I shall mark in white, at time t_2.

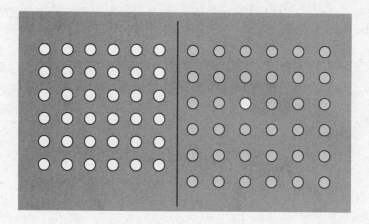

To see what the evolution of the universe would look like from this galaxy's vantage point, I simply superimpose the right image on the left, placing the galaxy in white on top of itself.

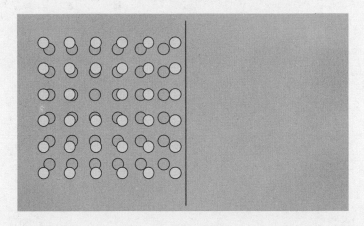

Voilà! From this galaxy's vantage point every other galaxy is moving away, and those that are twice as far have moved twice the distance in the same time, those that are three times as far away have moved three times the distance, etc. As long as there is no edge, those on the galaxy feel as if they are at the center of the expansion.

It doesn't matter what galaxy one chooses. Pick another galaxy, and repeat:

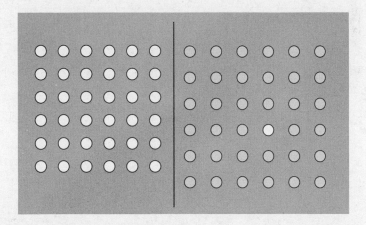

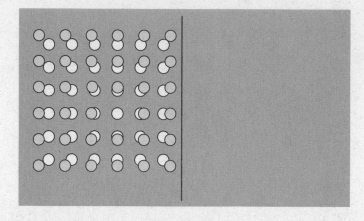

Depending upon your perspective, then, either *every place* is the center of the universe, or *no place* is. It doesn't matter; Hubble's law is consistent with a universe that is expanding.

Now, when Hubble and Humason first reported their analysis in 1929, they not only reported a linear relationship between distance and recession velocity, but also gave a quantitative estimate of the expansion rate itself. Here are the actual data presented at the time:

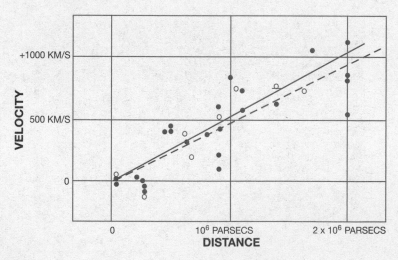

As you can see, Hubble's guess of fitting a straight line to this data set seems a relatively lucky one. (There is clearly some relationship, but whether a straight line is the best fit is far from clear on the basis of this data alone.) The number for the expansion rate they obtained, derived for the plot, suggested that a galaxy a million parsecs away (3 million light-years)—the average separation between galaxies—is moving away from us with a speed of 500 kilometers/second. This estimate was not so lucky, however.

The reason for this is relatively simple to see. If everything is moving apart today, then at earlier times they were closer together. Now, if gravity is an attractive force, then it should be slowing the expansion of the universe. This means the galaxy we see moving away from us at 500 kilometers/second today would have been moving faster earlier.

If for the moment, though, we just assume that the galaxy had always been carried away with that velocity, we can work backward and figure out how long ago it would have been at the same position as our galaxy. Since galaxies twice as far away are moving twice as fast, if we work backward we find out that they were superimposed on our position at exactly the same time. Indeed, the entire observable universe would have been superimposed at a single point, the Big Bang, at a time that we can estimate in this way.

Such an estimate is clearly an upper limit on the age of the universe, because, if the galaxies were once moving faster, they would have gotten where they are today in less time than this estimate would suggest.

From this estimate based on Hubble's analysis, the Big Bang happened approximately 1.5 billion years ago. Even in 1929, however, the evidence was already clear (except to some scriptural literalists in Tennessee, Ohio, and a few other states) that the Earth was older than 3 billion years old.

Now, it is embarrassing for scientists to find that the Earth is older than the universe. More important, it suggests something is wrong with the analysis.

The source of this confusion was simply the fact that Hubble's distance estimates, derived using the Cepheid relations in our galaxy, were systematically incorrect. The distance ladder based on using nearby Cepheids to estimate the distance of farther away Cepheids, and then to estimate the distance to galaxies in which yet more distant Cepheids were observed, was flawed.

The history of how these systematic effects have been overcome is too long and convoluted to describe here and, in any case, no longer matters because we now have a much better distance estimator.

One of my favorite Hubble Space Telescope photographs is shown below:

It shows a beautiful spiral galaxy far far away, long long ago (long long ago because the light from the galaxy takes some time—more than 50 million years—to reach us). A spiral galaxy such as this, which resembles our own, has about 100 billion stars within it. The bright core at its center contains perhaps 10 billion stars. Notice the star on the lower left corner that is shining with a brightness almost equal to these 10 billion stars. On first sighting it, you might reasonably assume that this is a much closer star in our own galaxy that got in the way of the picture. But in fact, it is a star in that same distant galaxy, more than 50 million light-years away.

Clearly, this is no ordinary star. It is a star that has just exploded, a supernova, one of the brightest fireworks displays in the universe. When a star explodes, it briefly (over the course of about a month or so) shines in visible light with a brightness of 10 billion stars.

Happily for us, stars don't explode that often, about once per hundred years per galaxy. But we are lucky that they do, because if they didn't, we wouldn't be here. One of the most poetic facts I know about the universe is that essentially every atom in your body was once inside a star that exploded. Moreover, the atoms in your left hand probably came from a different star than did those in your right. We are all, literally, star children, and our bodies made of stardust.

How do we know this? Well, we can extrapolate our picture of the Big Bang back to a time when the universe was about 1 second old, and we calculate that all observed matter was compressed in a dense plasma whose temperature should have been about 10 billion degrees (Kelvin scale). At this temperature nuclear reactions can readily take place between protons and neutrons as they bind together and then break apart from further collisions. Following this process as the universe cools, we can predict how frequently these primeval nuclear constituents will bind to form the nuclei of atoms heavier than hydrogen (i.e., helium, lithium, and so on).

When we do so, we find that essentially no nuclei—beyond lithium, the third lightest nucleus in nature—formed during the primeval fireball that was the Big Bang. We are confident that our calculations are correct because our predictions for the cosmic abundances of the lightest elements agree bang-on with these observations. The abundances of these lightest elements— hydrogen, deuterium (the nucleus of heavy hydrogen), helium, and lithium—vary by 10 orders of magnitude (roughly 25 percent of the protons and neutrons, by mass, end up in helium, while 1 in every 10 billion neutrons and protons ends up within a lithium nucleus). Over this incredible range, observations and theoretical predictions agree.

This is one of the most famous, significant, and successful predictions telling us the Big Bang really happened. *Only a hot Big Bang can produce the observed abundance of light elements and maintain consistency with the current observed expansion of the universe.* I carry a wallet card in my back pocket showing the comparison of the predictions of the abundance of light elements and the observed abundance so that, each time I meet someone who doesn't believe that the Big Bang happened, I can show it to them. I usually never get that far in my discussion, of course, because data rarely impress people who have decided in advance that something is wrong with the picture. I carry the card anyway and reproduce it for you later in the book.

While lithium is important for some people, far more important to the rest of us are all the heavier nuclei like carbon, nitrogen, oxygen, iron, and so on. These were *not* made in the Big Bang. The only place they can be made is in the fiery cores of stars. And the only way they could get into your body today is if these stars were kind enough to have exploded, spewing their products into the cosmos so they could one day coalesce in and around a small blue planet located near the star we call the Sun. Over the course of the history of our galaxy, about 200 million stars have exploded. These myriad stars sacrificed themselves, if you wish, so that one day you could be born. I suppose that qualifies them as much as anything else for the role of saviors.

It turns out a certain type of exploding star, called a Type 1a supernova, has been shown by careful studies performed over the 1990s to have a remarkable property: with high accuracy, those Type 1a supernovae that are intrinsically brighter also shine longer. The correlation, while not fully understood theoretically, is empirically very tight. This means that these supernovae are very good "standard candles." By this we mean that these supernovae can be used to calibrate distances because their intrinsic brightness can be directly ascertained by a measurement that is independent of their distance. If we observe a supernova in a distant galaxy — and we can because they are very bright — then by observing how long it shines, we can infer its intrinsic brightness. Then, by measuring its apparent brightness with our telescopes, we can accurately infer just how far away the supernova and its host galaxy are. Then, by measuring the "redshift" of the light from the stars in the galaxy, we can determine its velocity, and thus can compare velocity with distance and infer the expansion rate of the universe.

So far so good, but if supernovae explode only once every hundred years or so per galaxy, how likely are we ever to be able to see one? After all, the last supernova in our own galaxy witnessed on Earth was seen by Johannes Kepler in 1604! Indeed, it is said that supernovae in our galaxy are observed only during the lifetimes of the greatest astronomers, and Kepler certainly fits the bill.

Starting out as a humble mathematics teacher in Austria, Kepler became assistant to the astronomer Tycho Brahe (who himself had observed an earlier supernova in our galaxy and was given an entire island by the king of Denmark in return), and using Brahe's data on planetary positions in the sky taken over more than a decade, Kepler derived his famous three laws of planetary motion early in the seventeenth century:

1. Planets move around the Sun in ellipses.
2. A *line* connecting a planet and the Sun sweeps out equal *areas* during equal intervals of time.
3. The *square* of the *orbital period* of a planet is directly *proportional* to the *cube* (3rd power) of the *semi-major axis* of its orbit (or, in other words, of the "semi-major axis" of the ellipse, half of the distance across the widest part of the ellipse).

These laws in turn lay the basis for Newton's derivation of the universal law of gravity almost a century later. Besides this remarkable contribution, Kepler successfully defended his mother in a witchcraft trial and wrote what was perhaps the first science fiction story, about a journey to the moon.

Nowadays, one way to see a supernova is simply to assign a different graduate student to each galaxy in the sky. After all, one hundred years is not too different, in a cosmic sense at least, from the average time to do a PhD, and graduate students are cheap and abundant. Happily, however, we don't have to resort to such extreme measures, for a very simple reason: the universe is big and old and, as a result, rare events happen all the time.

Go out some night into the woods or desert where you can see stars and hold up your hand to the sky, making a tiny circle between your thumb and forefinger about the size of a dime.

Hold it up to a dark patch of the sky where there are no visible stars. In that dark patch, with a large enough telescope of the type we now have in service today, you could discern perhaps 100,000 galaxies, each containing billions of stars. Since supernovae explode once per hundred years per galaxy, with 100,000 galaxies in view, you should expect to see, on average, about three stars explode on a given night.

Astronomers do just this. They apply for telescope time, and some nights they might see one star explode, some nights two, and some nights it might be cloudy and they might not see any. In this way several groups have been able to determine Hubble's constant with an uncertainty of less than 10 percent. The new number, about 70 kilometers per second for galaxies on average of 3 million light-years apart, is almost a factor of 10 smaller than that derived by Hubble and Humason. As a result, we infer an age of the universe of closer to 13 billion years, rather than 1.5 billion years.

As I shall describe later, this too is in complete agreement with independent estimates of the age of the oldest stars in our galaxy. From Brahe to Kepler, from Lemaître to Einstein and Hubble, and from the spectra of stars to the abundance of light elements, four hundred years of modern science have produced a remarkable and consistent picture of the expanding universe. Everything holds together. The Big Bang picture is in good shape.

CHAPTER 2

A Cosmic Mystery Story: Weighing the Universe

> *There are known knowns. These are things we know that we know. There are known unknowns. That is to say, there are things that we know we don't know. But there are also unknown unknowns. There are things we don't know we don't know.*
>
> —Donald Rumsfeld

Having established that the universe had a beginning, and that that beginning was a finite and measurable time in the past, a natural next question to ask is, "How will it end?"

In fact, this was the very question that led me to move from my home territory, particle physics, into cosmology. During the 1970s and 1980s, it became increasingly clear from detailed measurements of the motion of stars and gas in our galaxy, as well as from the motion of galaxies in large groups of galaxies called clusters, that there was much more to the universe than meets either the eye or the telescope.

Gravity is the chief force operating on the enormous scale of galaxies, so measuring the motion of objects on these scales allows us to probe the gravitational attraction that drives this motion. Such measurements took off with the pioneering work of the American astronomer Vera Rubin and her colleagues in the early 1970s. Rubin had graduated with her doctorate from Georgetown after taking night classes while her husband waited in the car because she didn't know how to drive. She had applied to Princeton, but that university didn't accept women into their graduate astronomy program until 1975. Rubin rose to become only the second woman ever to be awarded the Gold Medal of the Royal Astronomical Society. That prize and her many other well-deserved honors stemmed from her groundbreaking measurements of the rotation rate of our galaxy. By observing stars and hot gas that were ever-farther from the center of our galaxy, Rubin determined that these regions were moving much faster than they should have been if the gravitational force driving their movement was due to the mass of all the observed objects within the galaxy. Due to her work, it eventually became clear to cosmologists that the only way to explain this motion was to posit the existence of significantly more mass in our galaxy than one could account for by adding up the mass of *all* of this hot gas and stars.

There was a problem, however, with this view. The very same calculations that so beautifully explain the observed abundance of the light elements (hydrogen, helium, and lithium) in the universe also tell us more or less how many protons and neutrons, the stuff of normal matter, must exist in the universe. This is because, like any cooking recipe—in this case nuclear cooking—the amount of your final product depends upon how much of each ingredient you start out with. If you double the recipe—four eggs instead of two, for example—you get more of the end product, in this case an omelet. Yet the initial density of protons and neutrons in the universe arising out of the Big Bang, as determined by fitting to the observed abundance of hydrogen, helium, and lithium, accounts for about twice the amount of material we can see in stars and hot gas. Where are those particles?

It is easy to imagine ways to hide protons and neutrons (snowballs, planets, cosmologists . . . none of them shines), so many physicists predicted that as many protons and neutrons lie in dark objects as visible objects. However, when we add up how much "dark matter" has to exist to explain the motion of material in our galaxy, we find that the ratio of total matter to visible matter is not 2 to 1, but closer to 10 to 1. If this is not a mistake, then the dark matter cannot be made of protons and neutrons. There are just not enough of them.

As a young elementary particle physicist in the early 1980s, learning of this possibility of the existence of exotic dark matter was extremely exciting to me. It implied, literally, that the dominant particles in the universe were not good old-fashioned garden-variety neutrons and protons, but possibly some new kind of elementary particle, something that didn't exist on Earth today, but something mysterious that flowed between and amidst the stars and silently ran the whole gravitational show we call a galaxy.

Even more exciting, at least for me, this implied three new lines of research that could fundamentally reilluminate the nature of reality.

1. If these particles were created in the Big Bang, like the light elements I have described, then we should be able to use ideas about the forces that govern the interactions of elementary particles (instead of the interactions of nuclei relevant to determine elemental abundance) to estimate the abundance of possible exotic new particles in the universe today.
2. It might be possible to derive the total abundance of dark matter in the universe on the basis of theoretical ideas in particle

physics, or it might be possible to propose new experiments to detect dark matter—either of which could tell us how much total matter there is and hence what the geometry of our universe is. The job of physics is not to invent things we cannot see to explain things we can see, but to figure out how to see what we cannot see—to see what was previously invisible, the known unknowns. Each new elementary particle candidate for dark matter suggests new possibilities for experiments to detect directly the dark matter particles parading throughout the galaxy by building devices on Earth to detect them as the Earth intercepts their motion through space. Instead of using telescopes to search for faraway objects, if the dark matter particles are in diffuse bunches permeating the entire galaxy, they are here with us now, and terrestrial detectors might reveal their presence.

3. If we could determine the nature of the dark matter, and its abundance, we might be able to determine how the universe will end.

This last possibility seemed the most exciting of all, so I will begin with it. Indeed, I got involved in cosmology because I wanted to be the first person to know how the universe would end.

It seemed like a good idea at the time.

When Einstein developed his theory of general relativity, at its heart was the possibility that space could curve in the presence of matter or energy. This theoretical idea became more than mere speculation in 1919 when two expeditions observed starlight curving around the Sun during a solar eclipse in precisely the degree to which Einstein had predicted should happen if the presence of the Sun curved the space around it. Einstein almost instantly became famous and a household name. (Most people today think it was the equation $E = mc^2$, which came fifteen years earlier, that did it, but it wasn't.)

Now, if space is potentially curved, then the geometry of our whole universe suddenly becomes a lot more interesting. Depending upon the total amount of matter in our universe, it could exist in one of three different types of geometries, so-called *open, closed,* or *flat.*

It is hard to envisage what a curved three-dimensional space might actually look like. Since we are three-dimensional beings, we can no more easily intuitively picture a curved three-dimensional space than the two-dimensional beings in the famous book *Flatland* could imagine what their world would look like to a three-dimensional observer if it were curved like the surface of a sphere. Moreover, if the curvature is very small, then it is hard to imagine how one might actually detect it in everyday life, just as, during the Middle Ages at least, some people felt the Earth must be flat because from their perspective it looked flat.

Curved three-dimensional universes are difficult to picture—a closed universe is like a three-dimensional sphere, which sounds pretty intimidating—but some aspects are easy to describe. If you looked far enough in any direction in a closed universe, you would see the back of your head.

A flat three-dimensional universe isn't flat like a pancake, but is actually the good old-fashioned universe you always thought you lived in—one where light travels in straight lines, and the three perpendicular axes x, y, and z, point in the same three directions everywhere throughout space (once you draw them emerging from some arbitrary point within space). In a curved space, light travels in curved trajectories and the three perpendicular axes, drawn at one point, end up pointing in different directions as you move about in space.

While these exotic geometries may seem amusing or impressive to talk about, operationally there is a much more important consequence of their existence. General relativity tells us unambiguously that a closed universe whose energy density is dominated by matter like stars and galaxies, and even more exotic dark matter, *must* one day recollapse in a process like the reverse of a Big Bang—a Big *Crunch,* if you will. An open universe will continue to expand forever at a finite rate, and a flat universe is just at the boundary, slowing down, but never quite stopping.

Determining the amount of dark matter, and thus the total density of mass in the universe, therefore promised to reveal the answer to the age-old question (at least as old as T. S. Eliot anyway): Will the universe end with a bang or a whimper? The saga of determining the total abundance of dark matter goes back at least a half century, and one could write a whole book about it, which in fact I have already done, in my book *Quintessence.* However, in this case, as I shall now demonstrate (with both words *and* then a picture), it is true that a single picture is worth at least a thousand (or perhaps a hundred thousand) words.

The largest gravitationally bound objects in the universe are called *superclusters* of galaxies. Such objects can contain thousands of individual galaxies or more and can stretch across tens of millions of light-years. Most galaxies exist in such superclusters, and indeed our own galaxy is located within the Virgo supercluster of galaxies, whose center is almost 60 million light-years away from us.

Since superclusters are so large and so massive, basically anything that falls into anything will fall into clusters. So if we could weigh superclusters of galaxies and then estimate the total density of such superclusters in the universe, we could then "weigh the universe," including all the dark matter. Then, using the equations of general relativity, we could determine whether there is enough matter to close the universe or not.

So far so good, but how can we weigh objects that are tens of millions of light-years across? Simple. Use gravity.

In 1936, Albert Einstein, following the urgings of an amateur astronomer, Rudi Mandl, published a short paper in the magazine *Science* titled "Lens-Like Action of a Star by the Deviation of Light in the Gravitational Field." In this brief note Einstein demonstrated the remarkable fact that space itself could act like a lens, bending light and magnifying it, just like the lenses in my own reading glasses.

It was a kindlier, gentler time in 1936, and it is interesting to read the informal beginning of Einstein's paper, which after all was published in a distinguished scientific journal: "Some time ago, R. W. Mandl paid me a visit and asked me to publish the results of a little calculation, which I had made at his request. This note complies with his wish." Perhaps this informality was accorded to him because he was Einstein, but I prefer to suppose that it was a product of the era, when scientific results were not yet always couched in language removed from common parlance.

In any case, the fact that light followed curved trajectories if space itself curved in the presence of matter was the first significant new prediction of general relativity and the discovery that led to Einstein's international fame, as I have mentioned. So it is perhaps not that surprising (as was recently discovered) that in 1912, well before Einstein had in fact even completed his general relativity theory, he had performed calculations—as he tried to find some observable phenomenon that would convince astronomers to test his ideas—that were essentially identical to those he published in 1936 at the request of Mr. Mandl. Perhaps because he reached the same conclusion in 1912 that he stated in his 1936 paper, namely "there is no great chance of observing this phenomenon," he never bothered to publish his earlier work. In fact, after examining his notebooks for both periods, we can't say for sure that he later even remembered having done the original calculations twenty-four years before.

What Einstein did recognize on both occasions is that the bending of light in a gravitational field could mean that, if a bright object was located well behind an intervening distribution of mass, light rays going out in various directions could bend around the intervening distribution and converge again, just as they do when they traverse a normal lens, producing either a magnification of the original object or the production of numerous image copies of the original object, some of which might be distorted (see figure below).

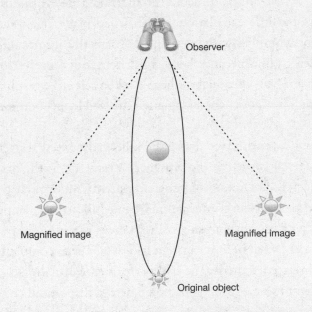

When he calculated the predicted effects for lensing of a distant star by an intervening star in the foreground, the effect was so small that it appeared absolutely unmeasurable, which led him to make the remark mentioned above—that it was unlikely that such a phenomenon could ever be observed. As a result, Einstein figured that his paper had little practical value. As he put it in his covering letter to the editor of *Science* at the time: "Let me also thank you for your cooperation with the little publication, which Mister Mandl squeezed out of me. It is of little value, but it makes the poor guy happy."

Einstein was not an astronomer, however, and it would take one to realize that the effect Einstein had predicted might be not only measurable, but also useful. Its usefulness came from applying it to the lensing of distant objects by much larger systems such as galaxies or even clusters of galaxies, not to the lensing of stars by stars. Within months of Einstein's publication, the brilliant Caltech astronomer Fritz Zwicky submitted a paper to the *Physical Review* in which he demonstrated the practicality of precisely this possibility (and also indirectly put down Einstein for his ignorance regarding the possible effect of lensing by galaxies rather than stars).

Zwicky was an irascible character and way ahead of his time. As early as 1933 he had analyzed the relative motion of galaxies in the Coma cluster and determined, using Newton's laws of motion, that the galaxies were moving so fast that they should have flown apart, destroying the cluster, unless there was far more mass in the cluster, by a factor more than 100, than could be accounted for by the stars alone. He thus should properly be considered as having discovered dark matter, though at the time his inference was so remarkable that most astronomers probably felt there might be some other less exotic explanation for the result he got.

Zwicky's one-page paper in 1937 was equally remarkable. He proposed three different uses for gravitational lensing: (1) testing general relativity, (2) using intervening galaxies as a kind of telescope to magnify more distant objects that would otherwise be invisible to telescopes on earth, and, most important, (3) resolving the mystery of why clusters appear to weigh more than can be accounted for by visible matter: *"Observations on the deflection of light around nebulae may provide the most direct determination of nebular masses and clear up the above-mentioned discrepancy."*

Zwicky's paper is now seventy-four years old but reads instead like a modern proposal for using gravitational lensing to probe the universe. Indeed, each and every suggestion he made has come to pass, and the final one is the most significant of all. Gravitational lensing of distant quasars by intervening galaxies was first observed in 1987, and in 1998, sixty-one years after Zwicky proposed weighing nebulae using gravitational lensing, the mass of a large cluster was determined by using gravitational lensing.

In that year, physicist Tony Tyson and colleagues at the now defunct Bell Laboratories (which had such a noble and Nobel tradition of great science, from the invention of the transistor to the discovery of the cosmic microwave background radiation) observed a distant large cluster, colorfully labeled CL 0024 + 1654, located about 5 billion light-years away. In this beautiful image from the Hubble Space Telescope, a spectacular example of the multiple image of a distant galaxy located another 5 billion light-years behind the cluster can be seen as highly distorted and elongated images amidst the otherwise generally rounder galaxies.

Looking at this image provides fuel for the imagination. First, every spot in the photo is a galaxy, not a star. Each galaxy contains perhaps 100 billion stars, along with them probably hundreds of billions of planets, and perhaps long-lost civilizations. I say long-lost because the image is 5 billion years old. The light was emitted 500 million years before our own Sun and Earth formed. Many of the stars in the photo no longer exist, having exhausted their nuclear fuel billions of years ago. Beyond that, the distorted images show precisely what Zwicky argued would be possible. The large distorted images to the left of the center of the image are highly magnified (and elongated) versions of this distant galaxy, which otherwise would probably not be visible at all.

Working backward from this image to determine the underlying mass distribution in the cluster is a complicated and complex mathematical challenge. To do so, Tyson had to build a computer model of the cluster and trace the rays from the source through the cluster in all possible different ways, using the laws of general relativity to determine the appropriate paths, until the fit they produced best matched the researchers' observations. When the dust settled, Tyson and collaborators obtained a graphical image that displayed precisely where the mass was located in this system pictured in the original photograph:

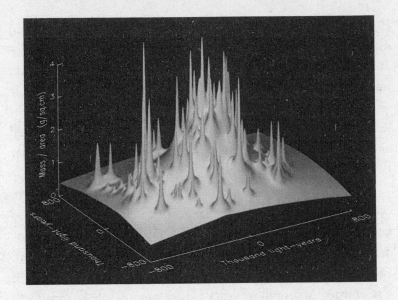

Something strange is going on in this image. The spikes in the graph represent the location of the visible galaxies in the original image, but most of the mass of the system is located *between* the galaxies, in a smooth, dark distribution. In fact, more than 40 times as much mass is between the galaxies as is contained in the visible matter in the system (300 times as much mass as contained in the stars alone with the rest of visible matter in hot gas around them). Dark matter is clearly not confined to galaxies, but also dominates the density of clusters of galaxies.

Particle physicists like myself were not surprised to find that dark matter also dominates clusters. Even though we didn't have a shred of direct evidence, we all hoped that the amount of dark matter was sufficient to result in a flat universe, which meant that there had to be more than 100 times as much dark matter as visible matter in the universe.

The reason was simple: a flat universe is the only mathematically beautiful universe. Why? Stay tuned.

Whether or not the total amount of dark matter was sufficient to produce a flat universe, observations such as these obtained by gravitational lensing (I remind you that gravitational lensing results from the local curvature of space around massive objects; the flatness of the universe relates to the global average curvature of space, ignoring the local ripples around massive objects) and more recent observations from other areas of astronomy have confirmed that the total amount of dark matter in galaxies and clusters is far in excess of that allowed by the calculations of Big Bang nucleosynthesis. We are now virtually certain that the dark matter—which, I reiterate, has been independently corroborated in a host of different astrophysical contexts, from galaxies to clusters of galaxies—must be made of something entirely new, something that doesn't exist normally on Earth. This kind of stuff, which isn't star stuff, isn't Earth stuff either. But it *is* something!

These earliest inferences of dark matter in our galaxy have spawned a whole new field of experimental physics, and I am happy to say that I have played a role in its development. As I have mentioned above, dark matter particles are all around us— in the room in which I am typing, as well as "out there" in space. Hence we can perform experiments to look for dark matter and for the new type of elementary particle or particles of which it is comprised.

The experiments are being performed in mines and tunnels deep underground. Why underground? Because on the surface of the Earth we are regularly bombarded by all manner of cosmic rays, from the Sun and from objects much farther away. Since dark matter, by its very nature, doesn't interact electromagnetically to produce light, we assume that its interactions with normal material are extremely weak, so it will be extremely difficult to detect. Even if we are bombarded every day by millions of dark matter particles, most will go through us and the Earth, without even "knowing" we are here—and without our noticing. Thus, if you want to detect the effects of the very rare exceptions to this rule, dark matter particles that actually bounce off atoms of matter, you had better be prepared to detect very rare and infrequent events. Only underground are you sufficiently shielded from cosmic rays for this to be possible even in principle.

As I write this, however, an equally exciting possibility is arising. The Large Hadron Collider, outside of Geneva, Switzerland, the world's largest and most powerful particle accelerator, has just begun running. But we have many reasons to believe that, at the very high energies at which protons are smashed together in the device, conditions similar to those in the very early universe will be re-created, albeit over only microscopically small regions. In such regions the same interactions that may have first produced what are now dark matter particles during the very early universe may now produce similar particles in the laboratory! There is thus a great race going on. Who will detect dark matter particles first: the experimenters deep underground or the experimentalists at the Large Hadron Collider? The good news is that, if either group wins the race, no one loses. We all win, by learning what the ultimate stuff of matter really is.

Even though the astrophysical measurements I described don't reveal the identity of dark matter, they do tell us how much of it exists. A final, direct determination of the total amount of matter in the universe came from the beautiful inferences of gravitational lensing measurements like the one I have described combined with other observations of X-ray emissions from clusters. Independent estimates of the clusters' total mass is possible because the temperature of the gas in clusters that are producing the X-rays is related to the total mass of the system in which they are emitted. The results were surprising, and as I have alluded, disappointing to many of us scientists. For when the dust had settled, literally and metaphorically, the total mass in and around galaxies and clusters was determined to be only about 30 percent of the total amount of mass needed to result in a flat universe today. (Note that this is more than 40 times as much mass as can be accounted for by visible matter, which therefore makes up less than 1 percent of the mass needed to make up a flat universe.)

Einstein would have been amazed that his "little publication" ultimately was far from useless. Supplemented by remarkable new experimental and observational tools that opened new windows on the cosmos, new theoretical developments that would have amazed and delighted him, and the discovery of dark matter that probably would have raised his blood pressure, Einstein's small step into the world of curved space had ultimately turned into to a giant leap. By the early 1990s, the holy grail of cosmology had apparently been achieved. Observations had determined that we live in an open universe, one that would therefore expand forever. Or had they?

CHAPTER 3

Light from the Beginning
of Time

As it was in the beginning, is now, and shall ever be.
—Gloria Patri

If you think about it, trying to determine the net curvature of
the universe by measuring the total mass contained within it and
then using the equations of general relativity to work backward
has huge potential problems. Inevitably, you have to wonder
whether matter is hidden in ways that we cannot uncover. For
instance, we can only probe for the existence of matter within
these systems using the gravitational dynamics of visible systems
like galaxies and clusters. If significant mass somehow resided
elsewhere, we'd miss it. It would be much better to measure the
geometry of the whole visible universe directly.

But how can you measure the three-dimensional geometry
of the whole visible universe? It's easier to start with a simpler
question: How would you determine if a two-dimensional object
like the Earth's surface was curved if you couldn't go around the
Earth or couldn't go above it in a satellite and look down?

First, you could ask a high school student, What is the sum
of the angles in a triangle? (Choose the high school carefully,

however . . . a European one is a good bet.) You would be told 180 degrees, because the student no doubt learned Euclidean geometry—the geometry associated with flat pieces of paper. On a curved two-dimensional surface like a globe, you can draw a triangle, the sum of whose angles is far greater than 180 degrees. For example, consider drawing a line along the equator, then making a right angle, going up to the North Pole, then another right angle back down to the equator, as shown below. Three times 90 is 270, far greater than 180 degrees. Voilà!

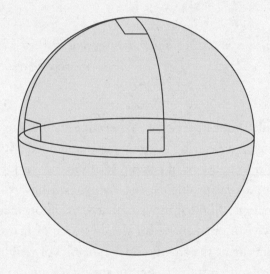

It turns out that this simple, two-dimensional thinking extends directly and identically to three dimensions, because the mathematicians who first proposed non-flat, or so-called non-Euclidean, geometries realized that the same possibilities could exist in three dimensions. In fact, the most famous mathematician of the nineteenth century, Carl Friedrich Gauss, was so fascinated by the possibility that our own universe might be curved that he took data in the 1820s and '30s from geodetic survey maps to measure large triangles between the German mountain peaks of Hoher Hagen, Inselberg, and Brocken to determine if he could detect any curvature of space itself. Of course, the fact that the mountains are on the curved surface of the Earth means that the two-dimensional curvature of the surface of the Earth would have interfered with any measurement he was performing to probe for curvature in the background three-dimensional space in which the Earth is situated, which he must have known. I assume he was planning to subtract any such contribution from his final results to see if any possible leftover curvature might be attributable to a curvature of the background space.

The first person to try to measure the curvature of space definitively was an obscure mathematician, Nikolai Ivanovich Lobachevsky, who lived in remote Kazan in Russia. Unlike Gauss, Lobachevsky was actually one of two mathematicians who had the temerity to propose in print the possibility of so-called hyberbolic curved geometries, where parallel lines could diverge. Remarkably, Lobachevsky published his work on hyperbolic geometry (which we now call "negatively curved" or "open" universes) in 1830.

Shortly thereafter, when considering whether our own three-dimensional universe might be hyperbolic, Lobachevsky suggested that it might be possible to "investigate a stellar triangle for an experimental resolution of the question." He suggested that observations of the bright star Sirius could be taken when the Earth was on either side of its orbit around the Sun, six months apart. From observations, he concluded that any curvature of our universe must be *at least* 166,000 times the radius of the Earth's orbit.

This is a big number, but it is trivially small on cosmic scales. Unfortunately, while Lobachevsky had the right idea, he was limited by the technology of his day. One hundred and fifty years later, however, things have improved, thanks to the most important set of observations in all of cosmology: measurements of the cosmic microwave background radiation, or CMBR.

The CMBR is nothing less than the afterglow of the Big Bang. It provides another piece of direct evidence, in case any is needed, that the Big Bang really happened, because it allows us to look back directly and detect the nature of the very young, hot universe from which all the structures we see today later emerged.

One of the many remarkable things about the cosmic microwave background radiation is that it was discovered in New Jersey, of all places, by two scientists who really didn't have the slightest idea what they were looking for. The other thing is that it existed virtually under all our noses for decades, potentially observable, but was missed entirely. In fact, you may be old enough have seen its effects without realizing it, if you remember the days before cable television, when channels used to end their broadcast days in the wee morning hours and not run infomercials all night. When they went off the air, after showing a test pattern, the screen would revert to static. About 1 percent of that static you saw on the television screen was radiation left over from the Big Bang.

The origin of the cosmic microwave background radiation is relatively straightforward. Since the universe has a finite age (recall it is 13.72 billion years old), and as we look out at ever more distant objects, we are looking further back in time (since the light takes longer to get to us from these objects), you might imagine that if we looked out far enough, we would see the Big Bang itself. In principle this is not impossible, but in practice, between us and that early time lies a wall. Not a physical wall like the walls of the room in which I am writing this, but one that, to a great extent, has the same effect.

I cannot see past the walls in my room because they are opaque. They absorb light. Now, as I look out in the sky back further and further in time, I am looking at the universe as it was younger and younger, and also hotter and hotter, because it has been cooling ever since the Big Bang. If I look back far enough, to a time when the universe was about 300,000 years old, the temperature of the universe was about 3,000 degrees (Kelvin scale) above absolute zero. At this temperature the ambient radiation was so energetic that it was able to break apart the dominant atoms in the universe, hydrogen atoms, into their separate constituents, protons and electrons. Before this time, neutral matter did not exist. Normal matter in the universe, made of atomic nuclei and electrons, consisted of a dense "plasma" of charged particles interacting with radiation.

A plasma, however, can be opaque to radiation. The charged particles within the plasma absorb photons and reemit them so that radiation cannot easily pass through such a material uninterrupted. As a result, if I try to look back in time, I cannot see past the time when matter in the universe was last largely comprised of such a plasma.

Once again, it is like the walls in my room. I can see them only because electrons in atoms on the surface of the wall absorb light from the light in my study and then reemit it, and the air between me and the walls is transparent, so I can see all the way to the surface of the wall that emitted the light. So too with the universe. When I look out, I can see all the way back to that "last scattering surface," which is the point at which the universe became neutral, where protons combined with electrons to form neutral hydrogen atoms. After that point, the universe became largely transparent to radiation, and I can now see the radiation that was absorbed and reemitted by the electrons as matter in the universe became neutral.

It is therefore a *prediction* of the Big Bang picture of the universe that there should be radiation coming at me from all directions from that "last scattering surface." Since the universe has expanded by a factor of about 1,000 since that time, the radiation has cooled on its way to us and is now approximately 3 degrees above absolute zero. And that is precisely the signal that the two hapless scientists found in New Jersey in 1965, and for whose discovery they were later awarded the Nobel Prize.

Actually a second Nobel Prize was given more recently for observations of the cosmic microwave background radiation, and for good reason. If we could take a photo of the surface of the last scattering surface, we would get a picture of the neonatal universe a mere 300,000 years into its existence. We could see all the structures that would one day collapse to form galaxies, stars, planets, aliens, and all the rest. Most important, these structures would have been unaffected by all the subsequent dynamical evolution that can obscure the underlying nature and origin of the first tiny primordial perturbations in matter and energy which were presumably created by exotic processes in the earliest moments of the Big Bang.

Most important for our purpose, however, on this surface there would be a characteristic scale, which is imprinted by nothing other than time itself. One can understand this as follows: If one considers a distance spanning about 1 degree on the last scattering surface as seen by an observer on Earth, this would correspond to a distance on that surface of about 300,000 light-years. Now, since the last scattering surface reflects a time when the universe itself was about 300,000 years old, and since

Einstein tells us that no information can travel through space faster than the speed of light, this means that no signal from one location could travel across this surface at that time by more than about 300,000 light-years.

Now consider a lump of matter smaller than 300,000 light-years across. Such a lump will have begun to collapse due to its own gravity. But a lump larger than 300,000 light-years across won't even begin to collapse, because it doesn't yet even "know" it is a lump. Gravity, which itself propagates at the speed of light, cannot have traveled across the full length of the lump. So just as Wile E. Coyote runs straight off a cliff and hangs suspended in midair in the Road Runner cartoons, the lump will just sit there, waiting to collapse when the universe becomes old enough for it to know what it is supposed to do!

This singles out a special triangle, with one side 300,000 light-years across, a known distance away from us, determined by the distance between us and the last scattering surface, as shown below:

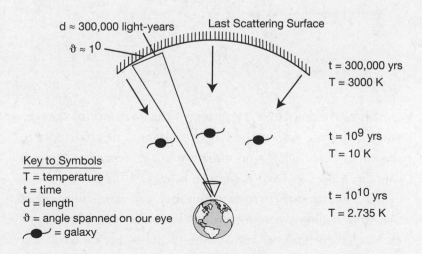

d ≈ 300,000 light-years   Last Scattering Surface

$\vartheta \approx 1^0$

t = 300,000 yrs
T = 3000 K

t = 10^9 yrs
T = 10 K

Key to Symbols
T = temperature
t = time
d = length
ϑ = angle spanned on our eye
= galaxy

t = 10^{10} yrs
T = 2.735 K

The largest lumps of matter, which will have already begun to collapse and in so doing will produce irregularities on the image of the microwave background surface, will span this angular scale. If we are able to obtain an image of this surface as it looked at that time, we would expect such hot spots to be, on average, the largest significant lumps we see in the image.

However, whether the angle spanned by this distance is precisely 1 degree will in fact be determined by the geometry of the universe. In a flat universe, light rays travel in straight lines. In an open universe, however, light rays bend outward as one follows them back in time. In a closed universe, light rays converge as one follows them backward. Thus, the actual angle spanned on our eyes by a ruler that is 300,000 light-years across, located at a distance associated with the last scattering surface, depends upon the geometry of the universe, as shown below:

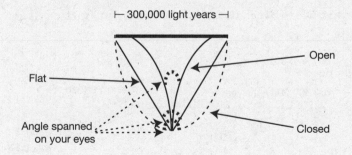

This provides a direct, clean test of the geometry of the universe. Since the size of the largest hot spots or cold spots in the microwave background radiation image depends just upon causality—the fact that gravity can propagate only at the speed of light, and thus the largest region that can have collapsed at that time is simply determined by the farthest distance a light ray can have propagated at that time—and because the angle that we see spanned by a fixed ruler at a fixed distance from us is just determined by the curvature of the universe, a simple picture of the last scattering surface can reveal to us the large-scale geometry of space-time.

The first experiment to attempt such an observation was a ground-launched balloon experiment in Antarctica in 1997 called BOOMERANG. While the acronym stands for *B*alloon *O*bservations *o*f *M*illimetric *E*xtragalactic *R*adiation *an*d *G*eophysics, the real reason it was called this name is simpler. A microwave radiometer was attached to a high-altitude balloon as shown below:

The balloon then went around the world, which is easy to do in Antarctica. Actually, at the South Pole it is really easy to do, since you can just turn around in a circle. However, from McMurdo Station the round trip around the continent, aided by the polar winds, took two weeks, after which the device returned to its starting point, hence the name BOOMERANG.

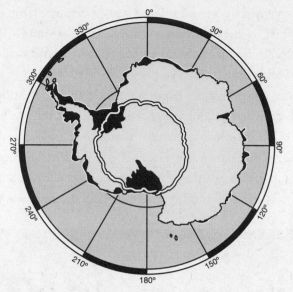

Boomerang path around Antarctica.

The purpose of the balloon trip was simple. To get a view of the microwave background radiation, reflecting a temperature of 3 degrees above absolute zero (Kelvin scale), which is not contaminated by the far hotter material on Earth (even in Antarctica temperatures are more than two hundred degrees hotter than the temperature of the cosmic microwave background radiation), we want to go as far as possible above the ground, and even above most of the atmosphere of the Earth. Ideally we use satellites for this purpose, but high-altitude balloons can do much of the job for far less money.

In any case, after two weeks, BOOMERANG returned an image of a small part of the microwave sky displaying hot and cold spots in the radiation pattern coming from the last scattering surface. Shown below is one image of the region the BOOMER-ANG experiment observed (with "hot spots" and "cold spots" being shaded dark and light respectively), superimposed upon the original photo of the experiment:

This image serves two purposes as far as I am concerned. First, it displays the actual physical scale of the hot and cold spots as seen in the sky by BOOMERANG, with the foreground images for comparison. But it also illustrates another important aspect of what can only be called our cosmic myopia. When we look up on a sunny day, we see a blue sky, as shown in the previous image of the balloon. But this is because we have evolved to see visible light. We have done so, no doubt, both because the light from the surface of our Sun peaks in the visible region, and also because many other wavelengths of light get absorbed in our atmosphere, so they cannot reach us on the Earth's surface. (This is fortunate for us, since much of this radiation could be harmful.) In any case, if we had instead evolved to "see" microwave radiation, the image of the sky we would see, day or night, as long as we weren't looking directly at the Sun, would take us directly back to an image of the last scattering surface, more than 13 billion light-years away. This is the "image" returned by the BOOMERANG detector.

The first flight of BOOMERANG, which produced this image, was remarkably fortunate. Antarctica is a hostile, unpredictable environment. On a later flight, in 2003, the entire experiment was nearly lost due to a balloon malfunction and subsequent storm. A last-minute decision to cut free from the balloon before it was blown to some inaccessible location saved the day and a search-and-rescue mission located the payload on the Antarctic plain and recovered the pressurized vessel containing the scientific data.

Before interpreting the BOOMERANG image, I want to emphasize one more time that the actual physical size of the hot spots and cold spots recorded on the BOOMERANG image are fixed by simple physics associated with the last scattering surface, while the *measured* sizes of the hot spots and cold spots in the image derive from the geometry of the universe. A simple two-dimensional analogy may help further explain the result: In two dimensions, a **closed geometry** resembles the surface of a sphere, while an **open geometry** resembles the surface of a saddle. If we draw a triangle on these surfaces, we observe the effect I described, as straight lines converge on a sphere, and diverge on a saddle, and, of course, remain straight on a flat plane:

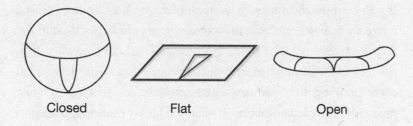

Closed   Flat   Open

So the million-dollar question now is, How big *are* the hot spots and cold spots in the BOOMERANG image? To answer this, the BOOMERANG collaboration prepared several simulated images on their computer of hot spots and cold spots as would be seen in closed, flat, and open universes, and compared this with (another false color) image of the actual microwave sky.

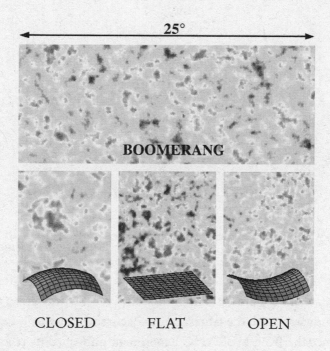

If you examine the image on the lower left, from a simulated closed universe, you will see that the average spots are larger than in the actual universe. On the right, the average spot size is smaller. But, just like Baby Bear's bed in Goldilocks, the image in the middle, corresponding to a simulated flat universe, is "just right." The mathematically beautiful universe hoped for by theorists seemed to be vindicated by this observation, even though it appears to conflict strongly with the estimate made by weighing clusters of galaxies.

In fact, the agreement between the predictions for a flat universe and the image obtained by BOOMERANG is almost embarrassing. Examining the spots and searching for the largest ones that had time to collapse significantly inward at the time reflected in the last scattering surface, the BOOMERANG team produced the following graph:

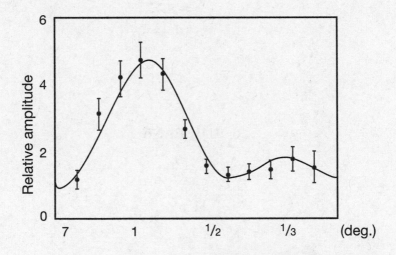

The data are the points. The solid line gives the prediction for a flat universe, with the largest bump occurring close to 1 degree!

Since the BOOMERANG experiment published its results, a far more sensitive satellite probe of the microwave background radiation was launched by NASA, the Wilkinson Microwave Anisotropy Probe (WMAP). Named after the late Princeton physicist David Wilkinson, who was one of the original Princeton physicists who should have discovered the CMBR had they not been scooped by the Bell Labs scientists, WMAP was launched in June 2001. It was sent out to a distance of one million miles from the Earth, where, on the far side of the Earth from the Sun, it could view the microwave sky without contamination from sunlight. Over a period of seven years it imaged the whole microwave sky (not just a portion of the sky as BOOMERANG did, since BOOMERANG had to contend with the presence of the Earth below it) with unprecedented accuracy.

Here the entire sky is projected on a plane, just as the surface of a globe can be projected on a flat map. The plane of our galaxy would lie along the equator, and 90 degrees above the plane of our galaxy is the North Pole on this map and 90 degrees below the plane of our galaxy is the South Pole. The image of the galaxy, however, has been removed from the map in order to reflect purely the radiation coming from the last scattering surface.

With this kind of exquisite data a much more precise estimate can be made of the geometry of the universe. A WMAP plot that is analogous to the one shown for the BOOMERANG image confirms to an accuracy of 1 percent that we live in a flat universe! The expectations of theorists were correct. Yet once again, we cannot ignore the apparent obvious inconsistency of this result with the result I described in the last chapter. Weighing the universe by measuring the mass of galaxies and clusters yields a value a factor of 3 smaller than the amount needed to result in a flat universe. Something has to give.

While theorists may have been patting themselves on the back for guessing that the universe is flat, almost no one was prepared for the surprise that nature had in store to resolve the contradictory estimates of the geometry of the universe coming from measuring mass versus measuring curvature directly. The missing energy needed to result in a flat universe turned out to be hiding right under our noses, literally.

CHAPTER 4

MUCH ADO ABOUT NOTHING

Less is more.
—LUDWIG MIES VAN DER ROHE,
AFTER ROBERT BROWNING

One step forward, two steps back, or so it seemed in our search for understanding our universe and accurately giving it a face. Even though observations had finally definitively determined the curvature of our universe—and in the process validated long-held theoretical suspicions—suddenly, even though it was known that ten times as much matter exists in the universe as could be accounted for by protons and neutrons, even that massive amount of dark matter, comprising 30 percent of what was required to produce a flat universe, was nowhere near sufficient to account for all the energy in the universe. The direct measurement of the geometry of the universe and the consequent discovery that the universe is indeed flat meant that 70 percent of the energy of the universe was still missing, neither in nor around galaxies or even clusters of galaxies!

Things were not quite as shocking as I have made them out to be. Even before these measurements of the curvature of the universe, and the determination of the total clustered mass within it (as described in chapter 2), there were signs that the by-then conventional theoretical picture of our universe—with sufficient dark matter (three times as much as we now know exists, in fact) to be spatially flat—was just not consistent with observations. Indeed, as early as 1995, I wrote a heretical paper with a colleague of mine, Michael Turner, from the University of Chicago, suggesting that this conventional picture couldn't be correct, and in fact the only possibility that appeared consistent with both a flat universe (our theoretical preference at the time) and observations of the clustering of galaxies and their internal dynamics was a universe that was far more bizarre and that hearkened back to a crazy theoretical idea Albert Einstein had in 1917 to solve the apparent contradiction between the predictions of his theory and the static universe he thought we lived in and which he later abandoned.

As I recall, our motivation at the time was more to show that something was wrong with the prevailing wisdom than it was to suggest a definitive solution to the problem. The proposal seemed too crazy to really believe, so I don't think anyone was more surprised than we were when it turned out, three years later, that our heretical suggestion was precisely on the money after all!

Let's return to 1917. Recall that Einstein had developed general relativity and had heart palpitations of joy over discovering that he could explain the precession of the perihelion of Mercury, even as he had to confront that fact that his theory couldn't explain the static universe in which he thought he was living.

Had he had greater courage of his convictions, he might have *predicted* that the universe couldn't be static. But he didn't. Instead, he realized that he could make a small change in his theory, one that was completely consistent with the mathematical arguments that had led him to develop general relativity in the first place, and one that looked like it might allow a static universe.

While the details are complex, the general structure of Einstein's equations in general relativity is relatively straightforward.

The left-hand side of the equations describes the curvature of the universe, and with it, the strength of the gravitational forces acting on matter and radiation. These are determined by the quantity on the right-hand side of the equation, which reflects the total density of all kinds of energy and matter within the universe.

Einstein realized that adding a small extra constant term to the left-hand side of the equation would represent a small extra constant *repulsive* force throughout all of space in addition to the standard gravitational attraction between distant objects that falls off as the distance between them increases. If it were small enough, this extra force could be undetectable on human scales or even on the scale of our solar system, where Newton's law of gravity is observed to hold so beautifully. But he reasoned that, because it was constant throughout all of space, it could build up over the scale of our galaxy and be large enough to counteract the attractive forces between very distant objects. He thus reasoned that this could result in a static universe on the largest scales.

Einstein called this extra term the *cosmological term*. Because it is simply a constant addition to the equations, it is now, however, conventional to call this term the *cosmological constant*.

Once he recognized that the universe is actually expanding, Einstein dispensed with this term and is said to have called the decision to add it to his equations his biggest blunder.

But getting rid of it is not so easy. It is like trying to put the toothpaste back in the tube after you have squeezed it out. This is because we now have a completely different picture of the cosmological constant today, so that, if Einstein had not added the term, someone else would have in the intervening years.

Moving Einstein's term from the left-hand side of his equations to the right-hand side is a small step for a mathematician but a giant leap for a physicist. While it is trivial mathematically to do so, once this term is on the right-hand side, where all the terms contributing to the energy of the universe reside, it represents something completely different from a physical perspective—namely a new contribution to the total energy. But what kind of stuff could contribute such a term?

The answer is, *nothing.*

By *nothing*, I do not mean nothing, but rather *nothing*—in this case, the nothingness we normally call empty space. That is to say, if I take a region of space and get rid of everything within it—dust, gas, people, and even the radiation passing through, namely absolutely *everything* within that region—if the remaining empty space *weighs something,* then that would correspond to the existence of a cosmological term such as Einstein invented.

Now, this makes Einstein's cosmological constant seem even crazier! For any fourth grader will tell you how much energy is contained in nothing, even if they don't know what energy is. The answer must be nothing.

Alas, most fourth graders have not taken quantum mechanics, nor have they studied relativity. For when one incorporates the results of Einstein's special theory of relativity into the quantum universe, empty space becomes much stranger than it was before. So strange in fact that even the physicists who first discovered and analyzed this new behavior were hard-pressed to believe that it actually existed in the real world.

The first person to successfully incorporate relativity into quantum mechanics was the brilliant, laconic British theoretical physicist Paul Dirac, who himself had already played a leading role in developing quantum mechanics as a theory.

Quantum mechanics was developed from 1912 to 1927, primarily through the work of the brilliant and iconic Danish physicist Niels Bohr and the brilliant young hot-shots Austrian physicist Erwin Schrödinger and German physicist Werner Heisenberg. The quantum world first proposed by Bohr, and refined mathematically by Schrödinger and Heisenberg, defies all common-sense notions based on our experience with objects on a human scale. Bohr first proposed that electrons in atoms orbit around the central nucleus, as planets do around the Sun, but demonstrated that the observed rules of atomic spectra (the frequencies of light emitted by different elements) could only be understood if somehow the electrons were restricted to have stable orbits in a fixed set of "quantum levels" and could not spiral freely toward the nucleus. They could move between levels by absorbing or emitting only discrete frequencies, or quanta, of light—the very quanta that Max Planck had first proposed in 1905 as a way of understanding the forms of radiation emitted by hot objects.

Bohr's "quantization rules" were rather ad hoc, however. In the 1920s, Schrödinger and Heisenberg independently demonstrated that it was possible to derive these rules from first principles if electrons obeyed rules of dynamics that were different from those applied to macroscopic objects like baseballs. Electrons could behave like waves as well as particles, appearing to spread out over space (hence, Schrödinger's "wave function" for electrons), and the results of measurements of the properties of electrons were shown to yield only probabilistic determinations, with various combinations of different properties not being exactly measurable at the same time (hence, Heisenberg's "Uncertainty Principle").

Dirac had shown that the mathematics proposed by Heisenberg to describe quantum systems (for which Heisenberg won the 1932 Nobel Prize) could be derived by careful analogy with the well-known laws governing the dynamics of classical macroscopic objects. In addition, he was also later able to show that the mathematical "wave mechanics" of Schrödinger could also be so derived and was formally equivalent to Heisenberg's formulation. But Dirac also knew that the quantum mechanics of Bohr, Heisenberg, and Schrödinger, as remarkable as it was, applied only to systems where Newton's laws, and not Einstein's relativity, would have been the appropriate laws governing the classical systems that the quantum systems were built with by analogy.

Dirac liked to think in terms of mathematics rather than pictures, and as he turned his attention to trying to make quantum mechanics consistent with Einstein's laws of relativity, he started playing with many different sorts of equations. These included complicated multicomponent mathematical systems that were necessary to incorporate the fact that electrons have "spin"—that is to say they spin around like small tops and have angular momentum, and they also can spin both clockwise and anticlockwise around any axis.

In 1929, he hit pay dirt. The Schrödinger equation had beautifully and accurately described the behavior of electrons moving at speeds much slower than light. Dirac found that if he modified the Schrödinger equation into a more complex equation using objects called matrices—which actually meant that his equation really described a set of four different coupled equations—he could consistently unify quantum mechanics with relativity, and thus in principle describe the behavior of systems where the electrons were moving at much faster speeds.

There was a problem, however. Dirac had written down an equation meant to describe the behavior of electrons as they interacted with electric and magnetic fields. But his equation appeared also to require the existence of new particles just like electrons but with opposite electric charge.

At the time, there was only one elementary particle in nature known with a charge opposite that of the electron—the proton. But protons are not at all like electrons. To begin with, they are 2,000 times heavier!

Dirac was flummoxed. In an act of desperation he argued that the new particles were in fact protons, but that somehow when moving through space the interactions of protons would cause them to act as if they were heavier. It didn't take long for others, including Heisenberg, to show that this suggestion made no sense.

Nature quickly came to the rescue. Within two years of the time Dirac proposed his equation, and a year after he had capitulated and accepted that, if his work was correct, then a new particle must exist, experimenters looking at cosmic rays bombarding the Earth discovered evidence for new particles identical to electrons but with an opposite electric charge, which were dubbed positrons.

Dirac was vindicated, but he also recognized his earlier lack of confidence in his own theory by later saying that his equation was smarter than he was!

We now call the positron the "antiparticle" of the electron, because it turns out that Dirac's discovery was ubiquitous. The same physics that required an antiparticle for the electron to exist requires one such particle to exist for almost every elementary particle in nature. Protons have antiprotons, for example. Even some neutral particles, like neutrons, have antiparticles. When particles and antiparticles meet, they annihilate into pure radiation.

While all this may sound like science fiction (and indeed antimatter plays an important role in *Star Trek*), we create antiparticles all the time at our large particle accelerators around the world. Because antiparticles otherwise have the same properties as particles, a world made of antimatter would behave the same way as a world of matter, with antilovers sitting in anticars making love under an anti-Moon. It merely is an accident of our circumstances, due, we think, to rather more profound factors we will get to later, that we live in a universe that is made up of matter and not antimatter or one with equal amounts of both. I like to say that while antimatter may seem strange, it is strange in the sense that Belgians are strange. They are not really strange; it is just that one rarely meets them.

The existence of antiparticles makes the observable world a much more interesting place, but it also turns out to make empty space much more complicated.

Legendary physicist Richard Feynman was the first person to provide an intuitive understanding of why relativity requires the existence of antiparticles, which also yielded a graphic demonstration that empty space is not quite so empty.

Feynman recognized that relativity tells us that observers moving at different speeds will make different measurements of quantities such as distance and time. For example, time will appear to slow down for objects moving very fast. If somehow objects could travel faster than light, they would appear to go backward in time, which is one of the reasons that the speed of light is normally considered a cosmic speed limit.

A key tenet of quantum mechanics, however, is the Heisenberg Uncertainty Principle, which, as I have mentioned, states that it is impossible to determine, for certain pairs of quantities, such as position and velocity, exact values for a given system at the same time. Alternatively, if you measure a given system for only a fixed, finite time interval, you cannot determine its total energy exactly.

What all this implies is that, for very short times, so short that you cannot measure their speed with high precision, quantum mechanics allows for the possibility that these particles act as if they are moving faster than light! But, if they are moving faster than light, Einstein tells us they must be behaving as if they are moving backward in time!

Feynman was brave enough to take this apparently crazy possibility seriously and explore its implications. He drew the following diagram for an electron moving about, periodically speeding up in the middle of its voyage to faster-than-light speed.

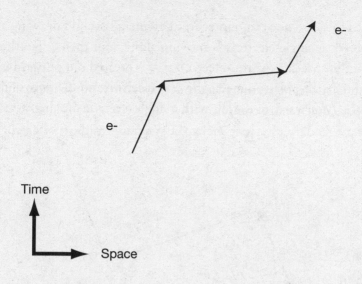

He recognized that relativity would tell us that another observer might alternatively measure something that would appear as shown below, with an electron moving forward in time, then backward in time, and then forward again.

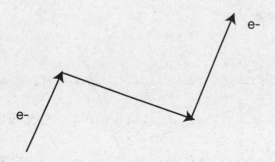

However, a negative charge moving backward in time is mathematically equivalent to a positive charge moving forward in time! Thus, relativity would require the existence of positively charged particles with the same mass and other properties as electrons.

In this case one can reinterpret Feynman's second drawing as follows: a single electron is moving along, and then at another point in space a positron-electron pair is created out of nothing, and then the positron meets the first electron and the two annihilate. Afterward, one is left with a single electron moving along.

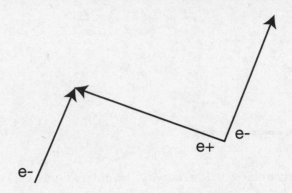

If this doesn't bother you, then consider the following: for a little while, even if you start out with just a single particle, and end with a single particle, for a short time there are three particles moving about:

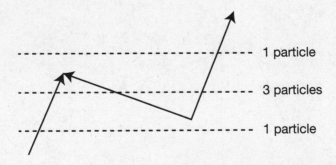

In the brief middle period, for at least a little while, something has spawned out of nothing! Feynman beautifully describes this apparent paradox in his 1949 paper, "A Theory of Positrons," with a delightful wartime analogy:

It is as though a bombardier watching a single road through the bomb-sight of a low-flying plane suddenly sees three roads and it is only when two of them come together and disappear again that he realizes that he has simply passed over a long switchback in a single road.

As long as this time period during this "switchback" is so short that we cannot measure all the particles directly, quantum mechanics and relativity imply that not only is this weird situation allowed, it is required. The particles that appear and disappear in timescales too short to measure are called *virtual* particles.

Now inventing a whole new set of particles in empty space that you cannot measure sounds a lot like proposing a large number of angels sitting on the head of a pin. And it would be about as impotent an idea if these particles had no other measurable effects. However, while they are not directly observable, it turns out their *indirect* effects produce most of the characteristics of the universe we experience today. Not only this, but one can calculate the impact of these particles more precisely than any other calculation in science.

Consider, for example, a hydrogen atom—the system Bohr tried to explain by developing his quantum theory and Schrödinger later tried to describe by deriving his famous equation. The beauty of quantum mechanics was that it could explain the specific colors of light emitted by hydrogen when it was heated up by arguing that electrons orbiting around the proton could exist only in discrete energy levels, and when they jumped between levels they absorbed or emitted only a fixed set of frequencies of light. The Schrödinger equation allows one to calculate the predicted frequencies, and it gets the answer almost exactly right.

But not exactly.

When the spectrum of hydrogen was observed more carefully, it was seen to be more complicated than had previously been estimated, with some additional small splittings between levels observed, called the "fine structure" of the spectrum. While these splittings had been known since Bohr's time, and it was suspected that perhaps relativistic effects had something do to with them, until a fully relativistic theory was available, no one could confirm this suspicion. Happily, Dirac's equation managed to improve the predictions compared to Schrödinger's equation and reproduced the general structure of the observations, including fine structure.

So far so good, but in April of 1947, United States experimentalist Willis Lamb and his student Robert C. Retherford performed an experiment that might otherwise seem incredibly ill motivated. They realized that they had the technological ability to measure the energy-level structure of the levels of hydrogen atoms with an accuracy of 1 part in 100 million.

Why would they bother? Well, whenever experimentalists find a new method to measure something with vastly greater precision than was possible before, that is often sufficient motivation for them to go ahead. Whole new worlds are often revealed in the process, as when the Dutch scientist Antonie Philips van Leeuwenhoek first stared at a drop of seemingly empty water with a microscope in 1676 and discovered it was teeming with life. In this case, however, the experimenters had more immediate motivation. Up until the time of Lamb's experiment, the available experimental precision could not test Dirac's prediction in detail.

The Dirac equation did predict the general structure of the new observations, but the key question that Lamb wanted to answer was whether it predicted it in detail. This was the only way to actually test the theory. And when Lamb tested the theory, it seemed to give the wrong answer, at a level of about 100 parts per billion, well above the sensitivity of his apparatus.

Such a small disagreement with experiment may not seem like a lot, but the predictions of the simplest interpretation of the Dirac theory were unambiguous, as was the experiment, and they differed.

Over the next few years, the best theoretical minds in physics jumped into the fray and tried to resolve the discrepancy. The answer came after a great deal of work, and when the dust had settled, it was realized that the Dirac equation actually gives precisely the correct answer, but only if you include the effect of virtual particles. Pictorially, this can be understood as follows. Hydrogen atoms are usually pictured in chemistry books something like this, with a proton at the center and an electron orbiting around it, jumping between different levels:

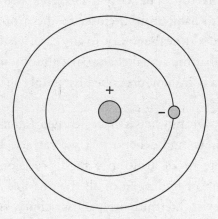

However, once we allow for the possibility that electron-positron pairs can spontaneously appear from nothing for a bit before annihilating each other again, over any short time the hydrogen atom really looks like this:

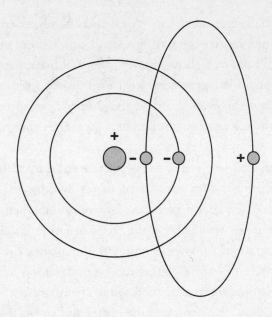

At the right of the figure I have drawn such a pair, which then annihilate at the top. The virtual electron, being negatively charged, likes to hang around closer to the proton, while the positron likes to stay farther away. In any case, what is clear from this picture is that the actual charge distribution in a hydrogen atom is *not*, at any instant, described by simply a single electron and proton.

Remarkably, we physicists have learned (after all the hard work by Feynman and others) that we can use Dirac's equation to calculate to an arbitrarily high precision, the impact on the spectrum of hydrogen of all the possible virtual particles that may exist intermittently in its vicinity. And when we do, we come up with *the best, most accurate prediction in all of science*. All other scientific predictions pale in comparison. In astronomy, the most recent observations of the cosmic microwave background radiation allow us to compare with theoretical predictions at the level of perhaps 1 part in 100,000, which is remarkable. However, using Dirac's equation, and the predicted existence of virtual particles, we can calculate the value of atomic parameters and compare them with observations and have remarkable agreement at the level of about 1 part in a billion or better!

Virtual particles therefore exist.

While the spectacular precision available in atomic physics is hard to match, there is nevertheless another place where virtual particles play a key role that may actually be more relevant to the central issue of this book. It turns out that they are responsible for most of your mass, and that of everything that is visible in the universe.

One of the great successes in the 1970s in our fundamental understanding of matter came with the discovery of a theory that accurately describes the interactions of quarks, the particles that make up the protons and neutrons that form the bulk of material from which you and everything you can see are made. The mathematics associated with the theory is complex, and it took several decades before techniques were developed that could handle it, particularly in the regime where the strong interaction between the quarks became appreciable. A herculean effort was launched, including building some of the most complicated parallel processing computers, which simultaneously utilize tens of thousands of individual processors, in order to try to calculate the fundamental properties of protons and neutrons, the particles we actually measure.

After all of this work, we now have a good picture of what the inside of a proton actually looks like. There may be three quarks contained therein, but there is also a lot of other stuff. In particular, virtual particles reflecting the particles and fields that convey the strong force between quarks are popping in and out of existence all the time. Here is a snapshot of how things actually look.

It is not a real photograph of course, but rather an artistic rendering of the mathematics governing the dynamics of quarks and the fields that bind them. The odd shapes and different shadings reflect the strength of the fields interacting with one another and with the quarks inside the proton as virtual particles spontaneously pop in and out of existence.

The proton is intermittently full of these virtual particles and, in fact, when we try to estimate how much they might contribute to the mass of the proton, we find that the quarks themselves provide very little of the total mass and that the fields created by these particles contribute most of the energy that goes into the proton's rest energy and, hence, its rest mass. The same is true for the neutron, and since you are made of protons and neutrons, the same is true for you!

Now, if we can calculate the effects of virtual particles on the otherwise empty space in and around atoms, and we can calculate the effects of virtual particles on the otherwise empty space inside of protons, then shouldn't we be able to calculate the effects of virtual particles on truly empty space?

Well, this calculation is actually harder to do. This is because, when we calculate the effect of virtual particles on atoms or on the proton mass, we are actually calculating the total energy of the atom or proton *including* virtual particles; then, we calculate the total energy that the virtual particles would contribute without the atom or proton present (i.e., in empty space); and *then* we subtract the two numbers in order to find the net impact upon the atom or proton. We do this because it turns out that each of these two energies is formally infinite when we attempt to solve the appropriate equations, but when we subtract the two quantities, we end up with a finite difference, and moreover one that agrees precisely with the measured value!

However, if we want to calculate the effect of virtual particles on empty space alone, we have nothing to subtract, and the answer we get is therefore infinite.

Infinity is not a pleasant quantity, however, at least as far as physicists are concerned, and we try to avoid it whenever possible. Clearly, the energy of empty space (or anything else, for that matter) cannot be physically infinite, so we have to figure out a way to do the calculation and get a finite answer.

The source of the infinity is easy to describe. When we consider all possible virtual particles that can appear, the Heisenberg Uncertainty Principle (which I remind you says that the uncertainty in the measured energy of a system is inversely proportional to the length of time over which you observe it) implies that particles carrying ever more energy can appear spontaneously out of nothing as long as they then disappear in ever-shorter times. In principle, particles can therefore carry almost infinite energy as long as they disappear in almost infinitesimally short times.

However, the laws of physics as we understand them apply only for distances and times larger than a certain value, corresponding to the scale where the effects of quantum mechanics must be considered when trying to understand gravity (and its associated effects on space-time). Until we have a theory of "quantum gravity," as it is called, we can't trust extrapolations that go beyond these limits.

Thus, we might hope that the new physics associated with quantum gravity will somehow cut off the effects of virtual particles that live for less time than the "Planck-time," as it is called. If we then consider the cumulative effects of only virtual particles of energies equal to or lower than that allowed by this temporal cutoff, we arrive at a finite estimate for the energy that virtual particles contribute to nothing.

But there is a problem. This estimate turns out to be about 1,000,000,000,000,000,000,000,000,000,000,000,000,000,000,000 ,000,000,000,000,000,000,000,000,000,000,000,000,000,000,000 ,000,000,000,000,000,000,000,000,000,000 times larger than the energy associated with all the known matter in the universe, including dark matter!

If the calculation of the atomic energy level spacings including virtual particles is the best computation in all of physics, this estimate of the energy of empty space—120 orders of magnitude larger than the energy of everything else in the universe—is undoubtedly the worst! If the energy of empty space were anywhere near this large, the repulsive force induced (remember the energy of empty space corresponds to a cosmological constant) would be large enough to blow up the Earth today, but more important, it would have been so great at early times that everything we now see in our universe would have pushed apart so quickly in the first fraction of a second of the Big Bang that no structure, no stars, no planets, and no people would ever have formed.

This problem, appropriately called the Cosmological Constant Problem, has been around since well before I was a graduate student, first made explicit by the Russian cosmologist Yakov Zel'dovich around 1967. It remains unsolved and is perhaps the most profound unsolved fundamental problem in physics today.

In spite of the fact that we have had no idea how to solve the problem for more than forty years, we theoretical physicists knew what the answer had to be. Like the fourth grader who I suggested would have guessed that the energy of empty space had to be zero, we too felt that when an ultimate theory was derived, it would explain how the effects of virtual particles would cancel, leaving empty space with precisely zero energy. Or nothing. Or rather, Nothing.

Our reasoning was better than the fourth grader's, or so we thought. We needed to reduce the magnitude of the energy of empty space from the truly gargantuan value that the naïve estimate suggested to a value consistent with the upper limits allowed by observation. This would require some way to subtract from a very large positive number another very large positive number so the two would cancel to 120 decimal places, leaving something non-zero in the 121st decimal place! But there is no precedent in science for canceling two large numbers to such accuracy, with only something minuscule left over.

However, zero is a number that is easy to produce. Symmetries of nature often allow us to demonstrate that there are precisely equal and opposite contributions coming from different parts of a calculation, canceling out exactly, with precisely nothing left over. Or, again, Nothing.

Thus, we theorists were able to rest easy and sleep at night. We didn't know how to get there, but we were sure what the final answer had to be.

Nature, however, had something different in mind.

CHAPTER 5

THE RUNAWAY UNIVERSE

It is mere rubbish, thinking at present of the origin of life;
one might as well think of the origin of matter.
—CHARLES DARWIN

What Michael Turner and I argued in 1995 was heretical in the extreme. Based on little more than theoretical prejudice, we presumed the universe was flat. (I should stress once again here that a "flat" three-dimensional universe is not flat like a two-dimensional pancake is flat, but is rather the three-dimensional space that all of us intuitively picture, in which light rays travel in straight lines. This is to be contrasted with the much harder to picture curved three-dimensional spaces in which light rays, which trace the underlying curvature of space, do not travel in straight lines.) Then we inferred that all available cosmological data at the time were consistent with a flat universe only if about 30 percent of the total energy resided in some form of "dark matter," as observations seemed to indicate existed around galaxies and clusters, but much more strangely than even this, that the remaining 70 percent of the total energy in the universe resided not in any form of matter, but rather in empty space itself.

Our idea was crazy by any standards. In order to result in a value for the cosmological constant consistent with our claim, the estimated value for this quantity described in the last chapter would have to be reduced somehow by 120 orders of magnitude and still not be precisely zero. This would involve the most severe fine-tuning of any physical quantity known in nature, without the slightest idea how to adjust it.

This was one of the reasons that, as I lectured at various universities about the quandary of a flat universe, I evoked mostly smiles and no more. I don't think many people took our proposal seriously, and I am not even sure Turner and I did. Our point in raising eyebrows with our paper was to illustrate graphically a fact that was beginning to dawn not just on us, but also on several of our theorist colleagues around the world: something looked wrong with the by-then "standard" picture of our universe, in which almost all the energy required by general relativity to produce a flat universe today was assumed to reside in exotic dark matter (with a pinch of baryons—i.e., us Earthlings, stars, visible galaxies—to salt the mix).

A colleague recently reminded me that for the two years following our modest proposal, it was referenced only a handful of times in subsequent papers, and apparently all but one or two of these were in papers written by Turner or me! As perplexing as our universe is, the bulk of the scientific community believed it couldn't be as crazy as Turner and I suggested it was.

The simplest alternative way out of the contradictions was the possibility that the universe wasn't flat but open (one in which parallel light rays today would curve apart if we traced their trajectory backward. This was of course before the cosmic microwave background measurements made it clear that this option was not viable.) However, even this possibility had its own problems, though the situation there remained far from clear as well.

Any high school physics student will happily tell you that gravity sucks—that is, it is universally attractive. Of course, like so many things in science, we now recognize that we have to expand our horizons because nature is more imaginative than we are. If for the moment we assume that the attractive nature of gravity implies that the expansion of the universe has been slowing down, recall that we get an upper limit on the age of the universe by assuming that the velocity of a galaxy located at a certain distance from us has been constant since the Big Bang. This is because, if the universe has been decelerating, then the galaxy was once moving away faster from us than it is now, and therefore it would have taken less time to get to its current position than if it had always been moving at its current speed. In an open universe dominated by matter, the deceleration of the universe would be slower than in a flat universe, and therefore the inferred age of the universe would be greater than it would be for a flat universe dominated by matter, for the same current measured expansion rate. It would in fact be much closer to the value we estimate by assuming a constant rate of expansion over cosmic time.

Remember that a non-zero energy of empty space would produce a cosmological constant—like gravitational repulsion—implying that the expansion of the universe would instead speed up over cosmic time, and therefore galaxies would previously have been moving apart more slowly than they are today. This would imply that it would have taken even longer to get to their present distance than it would for a constant expansion. Indeed, for a given measurement of the Hubble constant today, the longest possible lifetime of our universe (about 20 billion years) is obtained by including the possibility of a cosmological constant along with the measured amount of visible and dark matter, if we are free to adjust its value along with the density of matter in the universe today.

In 1996, I worked with Brian Chaboyer and our collaborators Pierre Demarque at Yale and postdoc Peter Kernan at Case
Western Reserve to put a lower limit on the age of the oldest stars
in our galaxy to be about 12 billion years. We did this by mod-
eling the evolution of millions of different stars on high-speed
computers and comparing their colors and brightness with actual
stars observed in globular clusters in our galaxy, which were long
thought to be among the oldest objects in the galaxy. Assuming about a billion years for our galaxy to form, this lower limit
effectively ruled out a flat universe dominated by matter and
favored one with a cosmological constant (one of the factors that
had weighed on the conclusions in my earlier paper with Turner),
while an open universe teetered on the hairy edge of viability.

However, the ages of the oldest stars involve inferences based
on observations at the edge of the then current sensitivity and,
in 1997, new observational data forced us to revise our estimates
downward by about 2 billion years, leading to a somewhat
younger universe. So the situation became much murkier, and all
three cosmologies once again appeared viable, sending many of
us back to the drawing board.

All of this changed in 1998, coincidentally the same year that
the BOOMERANG experiment demonstrated that the universe
is flat.

In the intervening seventy years since Edwin Hubble measured the expansion rate of the universe, astronomers had worked
harder and harder to pin down its value. Recall that in the 1990s
they had finally found a "standard candle"—that is, an object
whose intrinsic luminosity observers felt they could independently ascertain, so that, when they measured its apparent luminosity, they could then infer its distance. The standard candle
seemed to be reliable and was one that could be observed across
the depths of space and time.

A certain type of exploding star called Type Ia supernova had recently been demonstrated to exhibit a relationship between brightness and longevity. Measuring how long a given Type Ia supernova remained bright required, for the first time, taking into account time dilation effects due to the expansion of the universe, which imply that the measured lifetime of such a supernova is actually longer than its actual lifetime in its rest frame. Nonetheless, we could infer the absolute brightness and measure its apparent brightness with telescopes and ultimately determine the distance to the host galaxy in which the supernova had exploded. Measuring the redshift of the galaxy at the same time allowed us to determine velocity. Combining the two allows us to measure, with increasing accuracy, the expansion rate of the universe.

Because supernovae are so bright, they provide not only a great tool to measure the Hubble constant, they also allow observers to look back to distances that are a significant fraction of the total age of the universe.

This offered a new and exciting possibility, which observers viewed as a much more exciting quarry: measuring how Hubble's constant changes over cosmic time.

Measuring how a constant is changing sounds like an oxymoron, and it would be except for the fact that we humans live such brief lives, at least on a cosmic scale. On a human timescale the expansion rate of the universe is indeed constant. However, as I have just described, the expansion rate of the universe will change over cosmic time due to the effects of gravity.

The astronomers reasoned that if they could measure the velocity and distance of supernovae located far away—across the far reaches of the visible universe—then they could measure the rate at which the expansion of the universe was slowing down (since everyone assumed the universe was acting sensibly, and the dominant gravitational force in the universe was attractive). This in turn they hoped would reveal whether the universe was open, closed, or flat, because the rate of slowing as a function of time is different for each geometry.

In 1996, I was spending six weeks visiting Lawrence Berkeley Laboratory, lecturing on cosmology and discussing various science projects with my colleagues there. I gave a talk about our claim that empty space might have energy, and afterward, Saul Perlmutter, a young physicist who was working on detecting distant supernovae, came up to me and said, "We will prove you wrong!"

Saul was referring to the following aspect of our suggestion of a flat universe, 70 percent of the energy of which should be contained in empty space. Recall that such energy would produce a cosmological constant, leading to a repulsive force that would then exist throughout all of space and that would dominate the expansion of the universe, causing its expansion to *speed up*, not slow down.

As I have described, if the expansion of the universe was speeding up over cosmic time, then the universe would be older today than we would otherwise infer had the expansion been slowing down. This would then imply that the lookback in time to galaxies with a given redshift would be longer than it would otherwise be. In turn, if they have been receding from us for a longer time, this would imply that the light from them originated from farther away. The supernovae in galaxies at some given measured redshift would then appear fainter to us than if the light originated closer. Schematically, if one was measuring velocity versus distance, the slope of the curve for relatively nearby galaxies would allow us to determine the expansion rate today, and then whether the curve bent upward or downward for distant supernovae would tell us whether the universe was speeding up or slowing down over cosmic time.

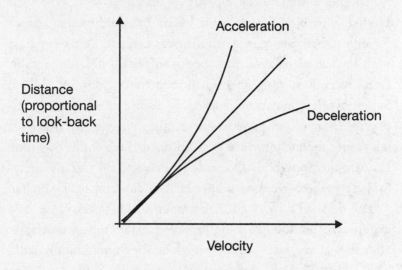

Two years after our meeting, Saul and his collaborators, part of an international team called the Supernova Cosmology Project, published a paper based on early preliminary data that indeed suggested we were wrong. (Actually, they did not argue that Turner and I were wrong, since they, along with most of the other observers, really didn't give much credence to our proposal.) Their data suggested that the distance-versus-redshift plot curved downward, and thus that an upper limit on the energy of empty space had to be well below what would have been required to make a significant contribution to the total energy today.

However, as often happens, the first data that come in might not be representative of all the data—either you are simply statistically unlucky, or unexpected systematic errors might affect the data, which are not manifest until you have a much bigger sample. This was the case with data that the Supernova Cosmology Project published, and so the conclusions were incorrect.

Another international supernova search project, called the High-Z Supernova Search Team, led by Brian Schmidt at Mount Stromlo Observatory in Australia, was carrying out a program with the same goal, and they began to obtain different results. Brian recently told me that when their first significant High-Z Supernova determination came in, suggesting an accelerating universe with significant vacuum energy, they were turned down for telescope time and informed by a journal that they must be wrong because the Supernova Cosmology Project had already determined that the universe was indeed flat, and dominated by matter.

The detailed history of the competition between these two groups will undoubtedly be replayed many times, especially after they share the Nobel Prize, which they undoubtedly will.* This is not the place to worry about priority. Suffice it to say that by early 1998, Schmidt's group published a paper demonstrating that the universe appeared to be accelerating. About six months later, Perlmutter's group announced similar results and published a paper confirming the High-Z Supernova result, in effect acknowledging their earlier error—and lending more credence to a universe dominated by the energy of empty space or, as it is now more commonly called, dark energy.

The speed with which these results were adopted by the scientific community—even though they required a wholesale revision of the entire accepted picture of the universe—provides an interesting study in scientific sociology. Almost overnight, there appeared to be universal acceptance of the results, even though, as Carl Sagan

---

* Indeed, as this book goes to print I just learned that Saul and Brian, along with Adam Reiss, who was part of the High-Z Supernova project, were awarded the Nobel Prize in Physics for 2011 for their discovery.

has emphasized, "Extraordinary claims require extraordinary evidence." This was certainly an extraordinary claim if ever one was.

I was shocked when, in December 1998, *Science* magazine called the discovery of an accelerating universe the Scientific Breakthrough of the Year, producing a remarkable cover with a drawing of a shocked Einstein.

I wasn't shocked because I thought that the result wasn't worthy of a cover. Quite the contrary. If true, it was one of the most important astronomical discoveries of our time, but the data at the time were merely strongly suggestive. They required such a change in our picture of the universe that I felt that we should all be more certain that other possible causes of the effects observed by the teams could be ruled out definitively before everyone jumped on the cosmological constant bandwagon. As I told at least one journalist at the time, "The first time I didn't believe in a cosmological constant was when observers claimed to discover it."

My somewhat facetious reaction may seem strange, given that I had been promoting the possibility in one form or another for perhaps a decade. As a theorist, I feel that speculation is fine, especially if it promotes new avenues for experiment. But I believe in being as conservative as possible when examining real data, perhaps because I reached scientific maturity during a period when so many new and exciting but tentative claims in my own field of particle physics turned out to be spurious. Discoveries ranging from a claimed new fifth force in nature to the discovery of new elementary particles to the supposed observation that our universe is rotating as a whole have come and gone with much hoopla.

The biggest concern at the time regarding the claimed discovery of an accelerating universe was that distant supernovae may appear dimmer than they would otherwise be expected to be, not because of an accelerated expansion, but merely because either (a) they *are* dimmer, or (b) perhaps some intergalactic or galactic dust present at early times partially obscures them.

In the intervening decade, it has nevertheless turned out that the evidence for acceleration has become overwhelming, almost unimpeachable. First, many more supernovae at high redshift have been measured. From these, a combined analysis of the supernovae from the two groups done within a year of the original publication yielded the following plot:

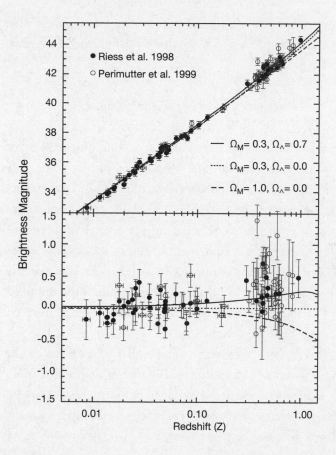

As a guide to the eye, to help you see whether the distance-versus-redshift curve bends upward or downward, the observers have drawn a dotted straight line in the upper half of the plot from the bottom left to the top right corner that goes through the data that represent nearby supernovae. The slope of this line tells us the expansion rate today. Then, in the lower half of the figure they have made that same straight line horizontal, to guide the eye. If the universe were decelerating, as had been expected in 1998, the distant supernovae at a redshift (z) close to 1 would fall below the straight line. But as you can see, most of them fall above the straight line. This is due to either one of two reasons:

1. the data are wrong, or
2. the expansion of the universe *is* accelerating.

If we take, for the moment, the second alternative and ask, "How much energy would we have to put in empty space in order to produce the observed acceleration?" the answer we come up with is remarkable. The solid curve, which fits the data best, corresponds to a flat universe, with 30 percent of the energy in matter and 70 percent in empty space. This is, remarkably, precisely what is needed in order to make a flat universe consistent with the fact that only 30 percent of the required mass exists in and around galaxies and clusters. An apparent concordance has been achieved.

Nevertheless, because the claim that 99 percent of the universe is invisible (1 percent visible matter embedded in a sea of dark matter surrounded by energy in empty space) fits into the category of an extraordinary claim, we should seriously consider the first of the two possibilities I mention above: namely, that the data are wrong. In the intervening decade, all the rest of the data from cosmology has continued to solidify the general concordance picture of a cockamamie, flat universe in which the dominant energy resides in empty space and in which everything we can see accounts for less than 1 percent of the total energy, with the matter we can't see being composed mostly of some yet unknown, new type of elementary particles.

First, new data on stellar evolution have improved as new satellites have provided us with information on the elemental abundances in old stars. Using these, my colleague Chaboyer and I were able, in 2005, to demonstrate definitively that the uncertainties in the estimates of the age of the universe using these data were now small enough to rule out lifetimes younger than about 11 billion years. This was inconsistent with any universe in which empty space itself did not contain a significant amount of energy. Again, since we are not certain that this energy is due to a cosmological constant, it now goes by the simpler name "dark energy," in analogue to the moniker of "dark matter" that dominates galaxies.

This estimate for the age of our universe was vastly improved in about 2006 when new precision measurements of the cosmic microwave background using the WMAP satellite allowed observers to precisely measure the time since the Big Bang. We now know the age of the universe to four significant figures. It is 13.72 billion years old!

I would never have figured that, in my lifetime, we would obtain such accuracy. But now that we have it, we can confirm that there is no way that a universe with the measured expansion rate today could be this old without dark energy, and in particular, dark energy that behaves essentially like the energy represented by a cosmological constant would behave. In other words, it is energy that appears to remain constant over time.

In the next scientific breakthrough, observers were able to measure accurately how matter, in the form of galaxies, has clustered together over cosmic time. The result depends upon the expansion rate of the universe, as the attractive force pulling galaxies together has to compete with the cosmic expansion driving matter apart. The larger the value of the energy of empty space, the sooner it will come to dominate the energy of the universe, and the sooner the increasing expansion rate will eventually stop the gravitational collapse of matter on ever larger scales.

By measuring gravitational clustering, therefore, observers have been able to confirm, once again, that the only flat universe that is consistent with observed large-scale structure in the universe is one with approximately 70 percent dark energy and, once again, that dark energy behaves more or less like the energy represented by a cosmological constant.

Independent of these indirect probes of the expansion history of the universe, the supernova observers have done extensive tests of possibilities that could induce systematic errors in their analysis, including the possibility of increased dust at large distances that make supernovae look dimmer, and ruled them out one by one.

One of their most important tests involved searching back in time.

Earlier in the history of the universe, when what is now our currently observable region was much smaller in size, the density of matter was much greater. However, the energy density of empty space remains the same over time if it reflects a cosmological constant—or something like it. Thus, when the universe was less than about half its present size, the energy density of matter would have exceeded the energy density of empty space. For all times before this time matter, and not empty space, would have produced the dominant gravitational force acting on the expansion. As a result, the universe would have been decelerating.

In classical mechanics there is a name for the point at which a system changes its acceleration and, in particular, goes from decelerating to accelerating. It is called a "jerk." In 2003, I organized a conference at my university to examine the future of cosmology and invited one of the High-Z Supernova survey members, Adam Riess, who had told me he would have something exciting to report at the meeting. He did. The next day, the *New York Times,* which was reporting on the meeting, ran a photo of Adam accompanied by the headline "Cosmic Jerk Discovered." I have kept that photo and turn to it for amusement from time to time.

The detailed mapping of the expansion history of the universe, demonstrating that it shifted from a period of deceleration to acceleration, added substantial weight to the claim that the original observations, which implied the existence of dark energy, were in fact correct. With all of the other evidence now available, it is very difficult to imagine that, by adhering to this picture, somehow we are being led on a cosmic wild-goose chase. Like it or not, dark energy seems here to stay, or at least to stay until it changes in some way.

The origin and nature of dark energy is without a doubt the biggest mystery in fundamental physics today. We have no deep understanding of how it originates and why it takes the value it has. We therefore have no idea of why it has begun to dominate the expansion of the universe and only relatively recently, in the past 5 billion years or so, or whether that is a complete accident. It is natural to suspect that its nature is tied in some basic way to the origin of the universe. And all signs suggest that it will determine the future of the universe as well.

CHAPTER 6

THE FREE LUNCH AT THE END
OF THE UNIVERSE

*Space is big. Really big. You just won't believe how vastly,
hugely, mind-bogglingly big it is. I mean, you may think
it's a long way down the road to the chemist's, but that's
just peanuts to space.*
—DOUGLAS ADAMS, *The Hitchhiker's Guide to the Galaxy*

One out of two isn't bad, I suppose. We cosmologists had
guessed, correctly it turned out, that the universe is flat, so we
weren't that embarrassed by the shocking revelation that empty
space indeed has energy—and enough energy in fact to dominate
the expansion of the universe. The existence of this energy was
implausible, but even more implausible is that the energy is not
enough to make the universe uninhabitable. For if the energy of
empty space were as large as the a priori estimates I described
earlier suggested it should be, the expansion rate would have
been so great that everything that we now see in the universe
would quickly have been driven beyond the horizon. The universe would become cold, dark, and empty well before stars, our
Sun, and our Earth could have formed.

Of all the reasons to suppose that the universe was flat, perhaps the simplest to understand arose from the fact that the universe had been well-known to be almost flat. Even in the early days, before dark matter was discovered, the known amount of visible material in and around galaxies accounted for perhaps 1 percent of the total amount of matter needed to result in a flat universe.

Now, 1 percent may not seem like much, but our universe is very old, billions of years old. Assuming that the gravitational effects of matter or radiation dominate the evolving expansion, which is what we physicists always thought was the case, then if the universe is not precisely flat, as it expands, it moves further and further away from being flat.

If it is open, the expansion rate continues at a faster rate than it would for a flat universe, driving matter farther and farther apart compared to what it would be otherwise, reducing its net density and very quickly yielding an infinitesimally small fraction of the density required to result in a flat universe.

If it is closed, then it slows the expansion down faster and eventually causes it to recollapse. All the while, the density first decreases at a slower rate than for a flat universe, and then as the universe recollapses, it starts to increase. Once again, the departure from the density expected for a flat universe increases with time.

The universe has increased in size by a factor of almost a trillion since it was 1 second old. If, at that earlier moment, the density of the universe was not almost exactly that expected of a flat universe but was, say, only 10 percent of that appropriate for a flat universe at the time, then today the density of our universe would differ from that of a flat universe by at least a factor of a trillion. This is far greater than the mere factor of 100 that was known to separate the density of the visible matter in the universe from the density of what would produce a flat universe today.

This problem was well-known, even in the 1970s, and it became known as the Flatness Problem. Considering the geometry of the universe is like imagining a pencil balancing vertically on its point on a table. The slightest imbalance one way or the other and it will quickly topple. So it is for a flat universe. The slightest departure from flatness quickly grows. Thus, how could the universe be so close to being flat today if it were not *exactly* flat?

The answer is simple: it must be essentially flat today!

That answer's actually not so simple, because it begs the question, How did initial conditions conspire to produce a flat universe?

There are two answers to this second, more difficult question. The first goes back to 1981, when a young theoretical physicist and postdoctoral researcher at Stanford University, Alan Guth, was thinking about the Flatness Problem and two other related problems with the standard Big Bang picture of the universe: the so-called Horizon Problem and the Monopole Problem. Only the first need concern us here, since the Monopole Problem merely exacerbates both the Flatness and Horizon problems.

The Horizon Problem relates to the fact that the cosmic microwave background radiation is extremely uniform. The small temperature deviations, which I described earlier, represented density variations in matter and radiation back when the universe was a few hundred thousand years old, of less than 1 part in 10,000 compared to the otherwise uniform background density and temperature. So while I was focused on the small deviations, a deeper, more urgent question was, How did the universe get to be so uniform in the first place?

After all, if instead of the earlier image of the CMBR (where temperature variations of a few parts in 100,000 are reflected in different colors), I showed a temperature map of the microwave sky on a linear scale (with variations in shades representing variations in temperature, of say, ±0.03 degree [Kelvin] about the mean background temperature of about 2.72 degrees above absolute zero, or a variation of 1 part in 100 about the mean), the map would look like this:

Compare this image, which contains nothing discernible in the way of structure, to a similar projection of the surface of the Earth, with only slightly greater sensitivity, with color variations representing variations about the mean radius by about 1 part in 500 or so:

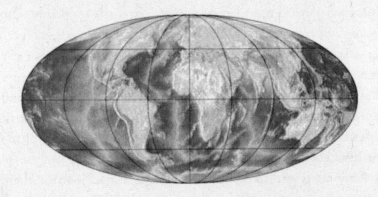

The universe is, therefore, on large scales, *incredibly uniform*! How could this be? Well, one might simply assume that, at early times, the early universe was hot, dense, and in thermal equilibrium. This means that any hot spots would have cooled, and cold spots would have heated up until the primordial soup reached the same temperature throughout.

However, as I pointed out earlier, when the universe was a few hundred thousand years old, light could have traveled only a few hundred thousand light-years, representing a small percentage of what is now the total observable universe (this former distance would represent merely an angle of about 1 degree on a map of the complete cosmic microwave background last scattering surface as it is observed today). Since Einstein tells us that no information can propagate faster than light, in the standard Big Bang picture, there is simply no way that one part of what is now the observable universe at that time would have been affected by the existence and temperature of other parts on angular scales of greater than about 1 degree. Thus, there is no way that the gas on these scales could have thermalized in time to produce such a uniform temperature throughout!

Guth, a particle physicist, was thinking about processes that could have occurred in the early universe that might have been relevant for understanding this problem when he came up with an absolutely brilliant realization. If, as the universe cooled, it underwent some kind of phase transition—as occurs, for example, when water freezes to ice or a bar of iron becomes magnetized as it cools—then not only could the Horizon Problem be solved, but also the Flatness Problem (and, for that matter, the Monopole Problem).

If you like to drink really cold beer, you may have had the following experience: you take a cold beer bottle out of the refrigerator, and when you open it and release the pressure inside the container, suddenly the beer freezes completely, during which it might even crack part of the bottle. This happens because, at high pressure, the preferred lowest energy state of the beer is in liquid form, whereas once the pressure has been released, the preferred lowest energy state of the beer is the solid state. During the phase transition, energy can be released because the lowest energy state in one phase can have lower energy than the lowest energy state in the other phase. When such energy is released, it is referred to as "latent heat."

Guth realized that, as the universe itself cooled with the Big Bang expansion, the configuration of matter and radiation in the expanding universe might have gotten "stuck" in some metastable state for a while until ultimately, as the universe cooled further, this configuration then suddenly underwent a phase transition to the energetically preferred ground state of matter and radiation. The energy stored in the "false vacuum" configuration of the universe before the phase transition completed—the "latent heat" of the universe, if you will—could dramatically affect the expansion of the universe during the period before the transition.

The false vacuum energy would behave just like that represented by a cosmological constant because it would act like an energy permeating empty space. This would cause the expansion of the universe at the time to speed up ever faster and faster. Eventually, what would become our observable universe would start to grow faster than the speed of light. This is allowed in general relativity, even though it seems to violate Einstein's special relativity, which says nothing can travel faster than the speed of light. But one has to be like a lawyer and parse this a little more carefully. Special relativity says nothing can travel *through space* faster than the speed of light. But *space itself* can do whatever the heck it wants, at least in general relativity. And as space expands, it can carry distant objects, which are at rest in the space where they are sitting, apart from one another at superluminal speeds.

It turns out that the universe could have expanded during this inflationary period by a factor of more than 10^{28}. While this is an incredible amount, it amazingly could have happened in a fraction of a second in the very early universe. In this case, everything within our entire observable universe was once, before inflation happened, contained in a region much smaller than we would have traced it back to if inflation had not happened, and most important, so small that there would have then been enough time for the entire region to thermalize and reach exactly the same temperature.

Inflation made another relatively generic prediction possible. When a balloon gets blown up larger and larger, the curvature at its surface gets smaller and smaller. Something similar happens for a universe whose size is expanding exponentially, as can occur during inflation—driven by a constant and large false vacuum energy. Indeed, by the time inflation ends (solving the Horizon Problem), the curvature of the universe (if it is non-zero to begin with) gets driven to an absurdly small value so that, even today, the universe appears essentially flat when measured accurately.

Inflation is the only currently viable explanation of both the homogeneity and flatness of the universe, based on what could be fundamental and calculable microscopic theories of particles and their interactions. But more than this, inflation makes another, perhaps even more remarkable prediction possible. As I have described already, the laws of quantum mechanics imply that, on very small scales, for very short times, empty space can appear to be a boiling, bubbling brew of virtual particles and fields wildly fluctuating in magnitude. These "quantum fluctuations" may be important for determining the character of protons and atoms, but generally they are invisible on larger scales, which is one of the reasons why they appear so unnatural to us.

However, during inflation, these quantum fluctuations can determine when what would otherwise be different small regions of space end their period of exponential expansion. As different regions stop inflating at slightly (microscopically) different times, the density of matter and radiation that results when the false vacuum energy gets released as heat energy in these different regions is slightly different in each one.

The pattern of density fluctuations that result after inflation—arising, I should stress, from the quantum fluctuations in otherwise empty space—turns out to be precisely in agreement with the observed pattern of cold spots and hot spots on large scales in the cosmic microwave background radiation. While consistency is not proof, of course, there is an increasing view among cosmologists that, once again, if it walks like a duck and looks like a duck and quacks like a duck, it is probably a duck. And if inflation indeed is responsible for all the small fluctuations in the density of matter and radiation that would later result in the gravitational collapse of matter into galaxies and stars and planets and people, then it can be truly said that we all are here today because of quantum fluctuations in what is essentially *nothing*.

This is so remarkable I want to stress it again. Quantum fluctuations, which otherwise would have been completely invisible, get frozen by inflation and emerge afterward as density fluctuations that produce everything we can see! If we are all stardust, as I have written, it is also true, if inflation happened, that we all, literally, emerged from quantum nothingness.

This is so strikingly nonintuitive that it can seem almost magical. But there is at least one aspect of all of this inflationary prestidigitation that might seem particularly worrisome. Where does all the energy come from in the first place? How can a microscopically small region end up as a universe-sized region today with enough matter and radiation within it to account for everything we can see?

More generally, we might ask the question, How is it that the density of energy can remain constant in an expanding universe with a cosmological constant, or false vacuum energy? After all, in such a universe, space expands exponentially, so that if the density of energy remains the same, the total energy within any region will grow as the volume of the region grows. What happened to the conservation of energy?

This is an example of something that Guth coined as the ultimate "free lunch." Including the effects of gravity in thinking about the universe allows objects to have—amazingly—"negative" as well as "positive" energy. This facet of gravity allows for the possibility that positive energy stuff, like matter and radiation, can be complemented by negative energy configurations that just balance the energy of the created positive energy stuff. In so doing, gravity can start out with an empty universe—and end up with a filled one.

This may also sound kind of fishy, but in fact it is a central part of the real fascination that many of us have with a flat universe. It is also something that you might be familiar with from high school physics.

Consider throwing a ball up in the air. Generally, it will come back down. Now throw it harder (assuming you are not indoors). It will travel higher and stay aloft longer before returning. Finally, if you throw it hard enough, it will not come down at all. It will escape the Earth's gravitational field and keep heading out into the cosmos.

How do we know when the ball will escape? We use a simple matter of energy accounting. A moving object in the gravitational field of the Earth has two kinds of energy. One, the energy of motion, is called *kinetic energy,* from the Greek word for motion. This energy, which depends upon the speed of the object, is always positive. The other component of the energy, called *potential energy* (related to the potential to do work), is generally negative.

This is the case because we define the total gravitational energy of an object located at rest infinitely far away from any other object as being zero, which seems reasonable. The kinetic energy is clearly zero, and we define the potential energy as zero at this point, so the total gravitational energy is zero.

Now, if the object is not infinitely far away from all other objects but is close to an object, like the Earth, it will begin to fall toward it because of the gravitational attraction. As it falls, it speeds up, and if it smacks into something on the way (say, your head), it can do work by, say, splitting it open. The closer it is to the Earth's surface when it is let go, the less work it can do by the time it hits the Earth. Thus, potential energy *decreases* as you get closer to the Earth. But if the potential energy is zero when it is infinitely far away from the Earth, it must get more and more negative the closer it gets to the Earth because its potential to do work decreases the closer it gets.

In classical mechanics, as I defined it here, the definition of potential energy is arbitrary. I could have set the potential energy of an object as zero at the Earth's surface, and then it would be some large number when the object is infinitely far away. Setting the total energy to zero at infinity does make physical sense, but it is, at least at this point in our discussion, merely a convention.

Regardless of where one sets the zero point of potential energy, the wonderful thing about objects that are subject to only the force of gravity is that the *sum* of their potential and kinetic energies remains a constant. As objects fall, potential energy is converted to the kinetic energy of motion, and as they bounce back up off the ground, kinetic energy is converted back to potential, and so on.

This allows us a marvelous bookkeeping tool to determine how fast one needs to throw something up in the air in order to escape the Earth, since if it eventually is to reach infinitely far away from the Earth, its total energy must be greater than or equal to zero. I then simply have to ensure that its total gravitational energy at the time it leaves my hand is greater than or equal to zero. Since I can control only one aspect of its total energy—namely the speed with which it leaves my hand—all I have to do is find the magic speed where the positive kinetic energy of the ball equals the negative potential energy it has due to the attraction at the Earth's surface. Both the kinetic energy and the potential energy of the ball depend precisely the same way on the mass of the ball, which therefore cancels out when these two quantities are equated, and one finds a single "escape velocity" for all objects from the Earth's surface, namely about 7 miles per second, when the total gravitational energy of the object is precisely zero.

What has all this got to do with the universe in general, and inflation in particular, you may ask? Well, the exact same calculation I just described for a ball that I throw up from my hand at the Earth's surface applies to every object in our expanding universe.

Consider a spherical region of our universe centered on our location (in the Milky Way galaxy) and large enough to encompass a lot of galaxies but small enough so that it is well within the largest distances we can observe today:

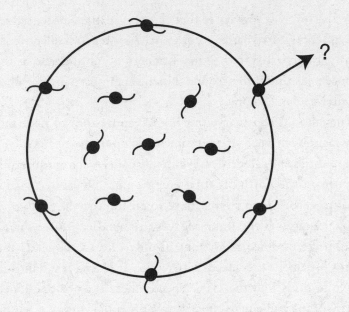

If the region is large enough but not too large, then the galaxies at the edge of the region will be receding from us uniformly due to the Hubble expansion, but their speeds will be far less than the speed of light. In this case, the laws of Newton apply, and we can ignore the effects of special and general relativity. In other words, every object is governed by physics that is identical to that which describes the balls that I have just imagined trying to eject from the Earth.

Consider the galaxy shown above, moving away from the center of the distribution as shown. Now, just as for the ball from the Earth, we can ask whether the galaxy will be able to escape from the gravitational pull of all the other galaxies within the sphere. And the calculation we would perform to determine the answer is precisely the same as the calculation we performed for the ball. We simply calculate the total gravitational energy of the galaxy, based on its motion outward (giving it positive energy), and the gravitational pull of its neighbors (providing a negative energy piece). If its total energy is greater than zero, it will escape to infinity, and if less than zero, it will stop and fall inward.

Now, remarkably, it is possible to show that we can rewrite the simple Newtonian equation for the total gravitational energy of this galaxy in a way that reproduces *exactly* Einstein's equation from general relativity for an expanding universe. And the term that corresponds to the total gravitational energy of the galaxy becomes, in general relativity, the term that describes the curvature of the universe.

So what do we then find? In a flat universe, and *only* in a flat universe, the total average Newtonian gravitational energy of each object moving with the expansion is *precisely zero*!

This is what makes a flat universe so special. In such a universe the positive energy of motion is exactly canceled by the negative energy of gravitational attraction.

When we begin to complicate things by allowing for empty space to have energy, the simple Newtonian analogy to a ball being thrown up in the air becomes incorrect, but the conclusion remains essentially the same. In a flat universe, even one with a small cosmological constant, as long as the scale is small enough that velocities are much less than the speed of light, the Newtonian gravitational energy associated with every object in the universe is zero.

In fact, with a vacuum energy, Guth's "free lunch" becomes even more dramatic. As each region of the universe expands to ever larger size, it becomes closer and closer to being flat, so that the total Newtonian gravitational energy of everything that results after the vacuum energy during inflation gets converted to matter and radiation becomes precisely zero.

But you can still ask, Where does all the energy come from to keep the density of energy constant during inflation, when the universe is expanding exponentially? Here, another remarkable aspect of general relativity does the trick. Not only can the gravitational energy of objects be negative, but their relativistic "pressure" can be negative.

Negative pressure is even harder to picture than negative energy. Gas, say in a balloon, exerts pressure on the walls of the balloon. In so doing, if it expands the walls of the balloon, it does work on the balloon. The work it does causes the gas to lose energy and cool. However, it turns out that the energy of empty space is gravitationally repulsive precisely because it causes empty space to have a "negative" pressure. As a result of this negative pressure, the universe actually does work *on* empty space as it expands. This work goes into maintaining the constant energy density of space even as the universe expands.

Thus, if the quantum properties of matter and radiation end up endowing even an infinitesimally small region of empty space with energy at very early times, this region can grow to be arbitrarily large and arbitrarily flat. When the inflation is over, one can end up with a universe full of stuff (matter and radiation), and the total Newtonian gravitational energy of that stuff will be as close as one can ever imagine to zero.

So when all the dust is settled, and after a century of trying, we have measured the curvature of the universe and found it to be zero. You can understand why so many theorists like me have found this not only very satisfying, but also highly suggestive.

A universe from Nothing . . . indeed.

CHAPTER 7

OUR MISERABLE FUTURE

The future ain't what it used to be.
—YOGI BERRA

In one sense it is both remarkable and exciting to find ourselves in a universe dominated by nothing. The structures we can see, like stars and galaxies, were all created by quantum fluctuations from nothing. And the average total Newtonian gravitational energy of each object in our universe is equal to nothing. Enjoy the thought while you can, if you are so inclined, because, if all this is true, we live in perhaps the worst of all universes one can live in, at least as far as the future of life is concerned.

Remember that barely a century ago, Einstein was first developing his general theory of relativity. Conventional wisdom then held that our universe was static and eternal. In fact, Einstein not only ridiculed Lemaître for suggesting a Big Bang, but also invented the cosmological constant for the purpose of allowing a static universe.

Now, a century later we scientists can feel smug for having discovered the underlying expansion of the universe, the cosmic microwave background, dark matter, and dark energy.

But what will the future bring?

Poetry . . . of a sort.

Recall that the domination of the expansion of our universe by the energy of seemingly empty space was inferred from the fact that this expansion is speeding up. And, just as with inflation, as described in the last chapter, our observable universe is at the threshold of expanding faster than the speed of light. And with time, because of the accelerated expansion, things will only get worse.

This means that, the longer we wait, the less we will be able to see. Galaxies that we can now see will one day in the future be receding away from us at faster-than-light speed, which means that they will become invisible to us. The light they emit will not be able to make progress against the expansion of space, and it will never again reach us. These galaxies will have disappeared from our horizon.

The way this works is a little different than you might imagine. The galaxies do not suddenly disappear or twinkle out of existence in the night sky. Rather, as their recession speed approaches the speed of light, the light from these objects gets ever more redshifted. Eventually, all their visible light moves to infrared, microwave, radio wave, and so on, until the wavelength of light they emit ends up becoming larger than the size of the visible universe, at which point they become officially invisible.

We can calculate about how long this will take. Since the galaxies in our local cluster of galaxies are all bound together by their mutual gravitational attraction, they will not recede with the background expansion of the universe discovered by Hubble. Galaxies just outside our group are about 1/5000th the distance out to the point where the recession velocity of objects approaches the speed of light. It will take them about 150 billion years, about 10 times the current age of the universe, to get there, at which point all the light from the stars within the galaxies will have redshifted by a factor of about 5,000. By about 2 trillion years, their light will have redshifted by an amount that will make their wavelength equal to the size of the visible universe, and the rest of the universe will literally have disappeared.

Two trillion years may seem like a long time, and it is. In a cosmic sense, however, it is nowhere near an eternity. The longest-lived "main sequence" stars (which have the same evolutionary history as our Sun) have lifetimes far longer than our Sun and will still be shining in 2 trillion years (even as our own Sun dies out in about only 5 billion years). And so in the far future there may be civilizations on planets around those stars, powered by solar power, with water and organic materials. And there may be astronomers with telescopes on those planets. But when they look out at the cosmos, essentially everything we can now see, all 400 billion galaxies currently inhabiting our visible universe, will have disappeared!

I have tried to use this argument with Congress to urge the funding of cosmology now, while we still have time to observe all that we can! For a congressperson, however, two years is a long time. Two trillion is unthinkable.

In any case, those astronomers in the far future would be in for a big surprise, if they had any idea what they were missing, which they won't. Because not only will the rest of the universe have disappeared, as my colleague Robert Scherrer of Vanderbilt and I recognized a few years ago, but essentially all of the evidence that now tells us we live in an expanding universe that began in a Big Bang will also have disappeared, along with all evidence of the existence of the dark energy in empty space that will be responsible for this disappearance.

While less than a century ago conventional wisdom still held that the universe was static and eternal, with stars and planets coming and going, but on its largest scales the universe itself perduring, in the far future, long after any remnants of our planet and civilization have likely receded into the dustbin of history, the illusion that sustained our civilization until 1930 will be an illusion that will once again return, with a vengeance.

There are three main observational pillars that have led to the empirical validation of the Big Bang, so that, even if Einstein and Lemaître had never lived, the recognition that the universe began in a hot, dense state would have been forced upon us: the observed Hubble expansion; the observation of the cosmic microwave background; and the observed agreement between the abundance of light elements—hydrogen, helium, and lithium—we have measured in the universe with the amounts predicted to have been produced during the first few minutes in the history of the universe.

Let's begin with the Hubble expansion. How do we know the universe is expanding? We measure the recession velocity of distant objects as a function of their distance. However, once all visible objects outside of our local cluster (in which we are gravitationally bound) have disappeared from our horizon, there will no longer be any tracers of the expansion—no stars, galaxies, quasars, or even large gas clouds—that observers could track. The expansion will be so efficient that it will have removed all objects from our sight that are actually receding from us.

Moreover, on a timescale of less than a trillion years or so, all the galaxies in our local group will have coalesced into some large meta-galaxy. Observers in the far future will see more or less precisely what observers in 1915 thought they saw: a single galaxy housing their star and their planet, surrounded by an otherwise vast, empty, static space.

Recall also that all evidence that empty space has energy comes from observing the rate of speed-up of our expanding universe. But, once again, without tracers of the expansion, the acceleration of our expanding universe will be unobservable. Indeed, in a strange coincidence, we are living in the only era in the history of the universe when the presence of the dark energy permeating empty space is likely to be detectable. It is true that this era is several hundred billion years long, but in an eternally expanding universe it represents the mere blink of a cosmic eye.

If we assume that the energy of empty space is roughly constant, as would be the case for a cosmological constant, then in much earlier times the energy density of matter and radiation would have far exceeded that in empty space. This is simply because, as the universe expands, the density of matter and radiation decreases along with the expansion because the distance between particles grows, so there are fewer objects in each volume. At earlier times, say earlier than about 5 billion to 10 billion years ago, the density of matter and radiation would have been far greater than it is today. The universe at this time and earlier was therefore dominated by matter and radiation, with their consequent gravitational attraction. In this case, the expansion of the universe would have been slowing down at these early times, and the gravitational impact of the energy of empty space would have been unobservable.

By the same token, far in the future, when the universe is several hundred billion years old, the density of matter and radiation will have decreased even further, and one can calculate that dark energy will have a mean energy density far in excess of a thousand billion times greater than the density of all remaining matter and radiation in the universe. It will, by then, completely govern the gravitational dynamics of the universe on large scales. However, at that late age, the accelerating expansion will have become essentially unobservable. In this sense, the energy of empty space ensures, by its very nature, that there is a finite time during which it is observable, and, remarkably, we live during this cosmological instant.

What about the other major pillar of the Big Bang, the cosmic microwave background radiation, which provides a direct baby picture of the universe? First, as the universe expands ever faster in the future, the temperature of the CMBR will fall. When the presently observable universe is about 100 times larger than it is now, the temperature of the CMBR will have fallen by a factor of 100, and its intensity, or the energy density stored within it, will have fallen by a factor of 100 million, making it about 100 million times harder to detect than it currently is.

But, after all, we have been able to detect the cosmic microwave background amidst all the other electronic noise on Earth, and we can imagine that observers in the far future will be 100 million times smarter than those we are blessed with today, so that all hope is not lost. Alas, it turns out that even the brightest observer one could imagine, with the most sensitive instrument one could build, will still be essentially out of luck in the distant future. This is because in our galaxy (or the meta-galaxy that will form when our galaxy merges with its neighbors, beginning with Andromeda in about 5 billion years) there is hot gas between stars, and this gas is ionized, so that it contains free electrons, and thus behaves like a plasma. As I described earlier, such a plasma is opaque to many types of radiation.

There is something called a "plasma frequency," below which radiation cannot permeate a plasma without absorption. Based on the currently observed density of free electrons in our galaxy, we can estimate the plasma frequency in our galaxy, and if we do this, we find that the bulk of the CMB radiation from the Big Bang will be stretched, by the time the universe gets to be about 50 times its present age, to long enough wavelengths, and hence low enough frequencies, that it will be below our future (meta-) galaxy's plasma frequency at that time. After that, the radiation will essentially not be able to make it into our (meta-)galaxy to be observed, no matter how tenacious the observer. The CMBR, too, will have disappeared.

So no observed expansion, no leftover afterglow of the Big Bang. But what about the abundance of the light elements— hydrogen, helium, and lithium—which also provides a direct signature of the Big Bang?

Indeed, as I described in chapter 1, whenever I meet someone who doesn't believe in the Big Bang, I like to show them the following figure that I keep as a card in my wallet. I then say: "See! There was a Big Bang!"

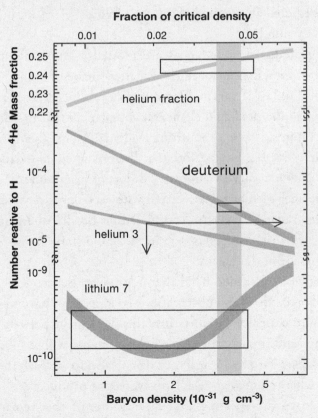

This figure looks very complicated, I know, but it actually shows the relative predicted abundance of helium, deuterium, helium-3, and lithium, compared to hydrogen, based on our current understanding of the Big Bang. The upper curve, going up and to the right, displays the predicted abundance of helium, the second most abundant element in the universe, by weight, compared with hydrogen (the most abundant element). The next two curves, going down and to the right represent the predicted abundances of deuterium and helium-3, respectively, not by weight but by number of atoms compared to hydrogen. Finally, the lower curve represents the predicted abundance of the next lightest element, lithium, again by number.

The predicted abundances are plotted as functions of the assumed total density of normal matter (made of atoms) in the universe today. If varying this quantity produced no combination of all the predicted elemental abundances that fit with our observations, it would be strong evidence against their production in a hot Big Bang. Note that the predicted abundances of these elements vary by almost 10 orders of magnitude.

The unshaded boxes associated with each curve represent the allowed range of the actual estimated primordial abundance of these elements based on observations of old stars and hot gas in and outside of our galaxy

The vertical shaded band then represents that region where all the predictions and observations *do* agree. It is hard to imagine more concrete support than this agreement between predictions and observations, again for elements whose predicted abundances vary by 10 orders of magnitude, for an early, hot Big Bang where all the light elements were first produced.

It is worth repeating the implications of this remarkable agreement more forcefully: Only in the first seconds of a hot

Big Bang, with an initial abundance of protons and neutrons that would result in something very close to the observed density of matter in visible galaxies today, and a density of radiation that would leave a remnant that would correspond precisely to the observed intensity of the cosmic microwave background radiation today, would nuclear reactions occur that could produce precisely the abundance of light elements, hydrogen and deuterium, helium and lithium, that we infer to have comprised the basic building blocks of the stars that now fill the night sky.

As Einstein might have put it, only a very malicious (and, therefore, in his mind unimaginable) God would have conspired to have created a universe that so unambiguously points to a Big Bang origin without its having occurred.

Indeed, when the rough agreement between the inferred helium abundance in the universe with the predicted helium abundance arising from a Big Bang was first demonstrated in the 1960s, this was one of the key bits of data that helped the Big Bang picture win out over the then very popular steady-state model of the universe championed by Fred Hoyle and his colleagues.

In the far future, however, things will be quite different. Stars burn hydrogen, producing helium, for example. At the present time only about 15 percent or so of all the observed helium in the universe could have been produced by stars in the time since the Big Bang—once again, a compelling bit of evidence that a Big Bang was required to produce what we see. But in the far future this will not be the case, because many more generations of stars will have lived and died.

When the universe is a trillion years old, for example, far more helium will have been produced in stars than will have been produced in the Big Bang itself. This situation is displayed in the following chart:

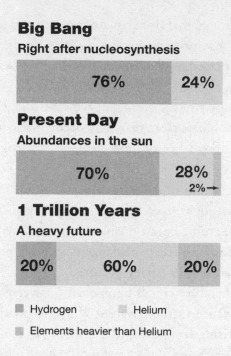

Big Bang
Right after nucleosynthesis

| 76% | 24% |

Present Day
Abundances in the sun

| 70% | 28% |

2%→

1 Trillion Years
A heavy future

| 20% | 60% | 20% |

■ Hydrogen   ■ Helium

■ Elements heavier than Helium

When 60 percent of the visible matter in the universe is comprised of helium, there will be no necessity for production of primordial helium in a hot Big Bang in order to produce agreement with observations.

Observers and theorists in some civilization in the far future will, however, be able to use this data to infer that the universe must have had a finite age. Because stars burn hydrogen to helium, there will be an upper limit on how long stars could have existed in order not to further deplete the ratio between hydrogen and helium. Thus, future scientists will estimate that the universe in which they live is less than about a trillion years old. But any direct signature that the beginning involved a Big Bang, rather than some other kind of spontaneous creation of our future single (meta-)galaxy, will be lacking.

Remember that Lemaître derived his claim of a Big Bang

purely on the basis of thinking about Einstein's general relativity. We can assume that any advanced civilization in the far future will discover the laws of physics, electromagnetism, quantum mechanics, and general relativity. Will some Lemaître of the far future therefore be able to derive a similar claim?

Lemaître's conclusion that our universe had to begin in a Big Bang was unavoidable, but it was based on an assumption that will not be true for the observable universe of the far future. A universe with matter stretching out uniformly in all directions, one that is isotropic and homogenous, cannot be static, for the reasons Lemaître and eventually Einstein recognized. However, there is a perfectly good solution of Einstein's equations for a single massive system surrounded by an otherwise empty static space. After all, if such a solution did not exist, then general relativity would not be able to describe isolated objects like neutron stars or, ultimately, black holes.

Large mass distributions like our galaxy are unstable, so eventually our (meta-)galaxy will itself collapse to form a massive black hole. This is described by a static solution of Einstein's equation called the Schwarzschild solution. But the time frame for our galaxy to collapse to form a massive black hole is much longer than the time frame for the rest of the universe to disappear. Thus, it will seem natural for scientists of the future to imagine that our galaxy could have existed for a trillion years in empty space without significant collapse and without requiring an expanding universe surrounding it.

Of course, speculations about the future are notoriously difficult. I am writing this, in fact, while at the World Economic Forum in Davos, Switzerland, which is full of economists who invariably predict the behavior of future markets and revise their predictions when they turn out to be horribly wrong. More generally, I find any predictions of the far future, and even the not-so-far future, of science and technology to be even sketchier than those of "the dismal science." Indeed, whenever I'm asked about the near future of science or what the next big breakthrough will be, I always respond that if I knew, I would be working on it right now!

Thus, I like to think of the picture I have presented in this chapter as something like the picture of the future presented by the third ghost in Dickens's *A Christmas Carol*. This is the future as it *might be*. After all, since we have no idea what the dark energy permeating empty space is, we also therefore cannot be certain that it will behave like Einstein's cosmological constant and remain constant. If it doesn't, the future of the universe could be far different. The expansion may not continue to accelerate, but instead may once again slow down over time so that distant galaxies will not disappear. Alternatively, perhaps there will be some new observable quantities we cannot yet detect that may provide astronomers in the future with evidence that there was once a Big Bang.

Nevertheless, based on everything we know about the universe today, the future I have sketched out is the most plausible one, and it is fascinating to consider whether logic, reason, and empirical data might still somehow induce future scientists to infer the correct underlying nature of our universe, or whether it will forever remain obscured behind the horizon. Some brilliant future scientist exploring the fundamental nature of forces and particles might derive a theoretical picture that will suggest that inflation must have happened, or that there must be an energy in empty space, which would further explain why there are no galaxies within the visible horizon. But I am not so sanguine about this.

Physics is, after all, an empirical science, driven by experiment and observation. Had we not observationally inferred the existence of dark energy, I doubt any theorist would have been bold enough to suggest its existence today. And while it is also possible to imagine tentative signatures that might suggest something is wrong with the picture of a single galaxy in a static universe without a Big Bang—perhaps some observation of elemental abundances that appears anomalous—I suspect that Occam's razor will suggest that the simplest picture is the correct one, and that the anomalous observations might be explained by some local effects.

Ever since Bob Scherrer and I laid out the challenge that future scientists will use falsifiable data and models—the very paragon of good science—but in the process that they will come up with a false picture of the universe, many of our colleagues have tried to suggest ways to probe that the universe is actually expanding in the far future. I too can imagine possible experiments. But I cannot see that they would be well motivated.

For example, you would need to eject bright stars from our galaxy and send them off into space, wait a billion years or so for them to explode, and try to observe their recession velocities as a function of the distance they reach before they explode in order to probe to see if they are getting any extra kick from a possible expansion of space. A tall order, but even if you could imagine somehow pulling this off, I cannot see the National Science Foundation of the future actually funding the experiment without at least some other motivation for arguing on behalf of an expanding universe. And if somehow stars from our galaxy are naturally ejected and detectable as they move out toward the horizon, it is not clear to me that observing an anomalous acceleration of some of these objects would be interpreted in terms of such a bold and strange proposal as an expanding universe dominated by dark energy.

We can consider ourselves lucky that we live at the present

time. Or as Bob and I put it in one of the articles we wrote: "We live at a very special time . . . the only time when we can observationally verify that we live at a very special time!"

We were being somewhat facetious, but it is sobering to suggest that one can use the best observational tools and theoretical tools at one's disposal and nevertheless come up with a completely false picture of the large-scale universe.

I should point out, nevertheless, that even though incomplete data *can* lead to a false picture, this is far different from the (false) picture obtained by those who choose to ignore empirical data to invent a picture of creation that would otherwise contradict the evidence of reality (young earthers, for example), or those who instead require the existence of something for which there is no observable evidence whatsoever (like divine intelligence) to reconcile their view of creation with their a priori prejudices, or worse still, those who cling to fairy tales about nature that presume the answers before questions can even be asked. At least the scientists of the future will be basing their estimates on the best evidence available to them, recognizing as we all do, or at least as scientists do, that new evidence may cause us to change our underlying picture of reality.

In this regard, it is worth adding that perhaps we are missing something even today that might have been observable had only we lived 10 billion years ago or perhaps could see if we lived 100 billion years into the future. Nevertheless, I should stress that the Big Bang picture is too firmly grounded in data from every area to be proved invalid in its general features. But some new, nuanced understanding of the fine details of the distant past or distant future, or of the origin of the Big Bang and its possible uniqueness in space, might easily emerge with new data. In fact, I hope it will. One lesson that we can draw from the possible future end of life and intelligence in the universe is that we need to have some cosmic humility in our claims, even if such a thing is difficult for cosmologists.

Either way, the scenario I have just described has a certain poetic symmetry, even if it is equally tragic. Long into the future, scientists will derive a picture of the universe that will hearken back to the very picture we had at the beginning of the last century, which itself ultimately served as the catalyst for investigations that led to the modern revolutions in cosmology. Cosmology will have come full circle. I for one find that remarkable, even if it underscores what some may view as the ultimate futility of our brief moment in the sun.

Regardless, the fundamental problem illustrated by the possible future end of cosmology is that we have only one universe to test—the one we live in. While test it we must if we want to have any hope of understanding how what we now observe arose, we nevertheless are limited in both what we can measure and in our interpretations of the data.

If many universes exist, and if we could somehow probe more than one, we might have a better chance of knowing which observations are truly significant and fundamental and which arise only as an accident of our circumstances.

As we shall see next, while the latter possibility is unlikely, the former is not, and scientists are pressing forward with new tests and new proposals to further our understanding of the unexpected and strange features of our universe.

Before proceeding, however, it is perhaps worth ending with another, more literary picture of the likely future I have presented here and one that is particularly relevant to the subject of this book. It comes from Christopher Hitchens's response to the scenario I have just described. As he put it, "For those who find it remarkable that we live in a universe of Something, just wait. Nothingness is heading on a collision course right toward us!"

CHAPTER 8

A Grand Accident?

*Once you assume a creator and a plan, it makes humans
objects in a cruel experiment whereby we are created to
be sick and commanded to be well.*

—Christopher Hitchens

We are hardwired to think that everything that happens to us
is significant and meaningful. We have a dream that a friend is
going to break her arm, and the next day we find out that she
sprained her ankle. Wow! Cosmic! Clairvoyant?

The physicist Richard Feynman used to like to go up to people and say: "You won't believe what happened to me today!
You just won't believe it!" And when they would inquire what
happened, he would say, "Absolutely nothing!" By this he was
suggesting that when something like the dream I described
above happens, people ascribe significance to it. But they forget
the myriad nonsense dreams they had that predicted absolutely
nothing. By forgetting that most of the time nothing of note
occurs during the day, we then misread the nature of probability when something unusual does occur: among any sufficiently
large number of events, something unusual is bound to happen
just by accident.

How does this apply to our universe?

Until the discovery that, inexplicably, the energy of empty space is not only not zero, but takes a value that is 120 orders of magnitude smaller than the estimate I described based on ideas from particle physics suggests, the conventional wisdom among physicists was that every fundamental parameter we measured in nature *is* significant. By this I mean that, somehow, on the basis of fundamental principles, we would eventually be able to understand things such as why gravity is so much weaker than the other forces of nature, why the proton is 2,000 times heavier than the electron, and why there are three families of elementary particles. Put another way, once we understood the fundamental laws that govern the forces of nature at its smallest scales, all of these current mysteries would be revealed as natural consequences of these laws.

(A purely religious argument, on the other hand, could take significance to an extreme by suggesting that each fundamental constant is significant because God presumably chose each one to have the value it does as part of a divine plan for our universe. In this case, nothing is an accident, but by the same token, nothing is predicted or actually explained. It is an argument by fiat that goes nowhere and yields nothing useful about the physical laws governing the universe, other than perhaps providing consolation for the believer.)

But the discovery that empty space has energy started a revision in thinking among many physicists about what is required in nature and what may be accidental.

The catalyst for this new gestalt originates from the argument I gave in the last chapter: dark energy is measurable today because "now" is the only time in the history of the universe when the energy in empty space is comparable to the energy density in matter.

Why should we be living at such a "special" time in the history of the universe? Indeed, this flies in the face of everything that has characterized science since Copernicus. We have learned that the Earth is not the center of the solar system and that the Sun is a star on the lonely outer edges of a galaxy that is merely one out of 400 billion galaxies in the observable universe. We have come to accept the "Copernican principle" that there is nothing special about our place and time in the universe.

But with the energy of empty space being what it is, we *do* appear to live at a special time. This is shown best by the following illustration of a "brief history of time."

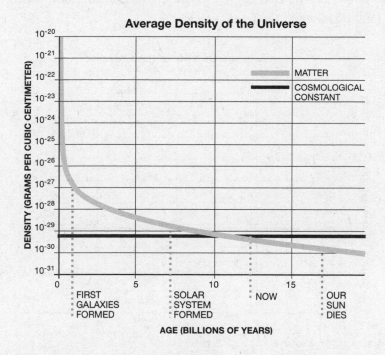

The two curves represent the energy density of all matter in the universe, and the energy density of empty space (presuming it is a cosmological constant) as a function of time. As you can see, the density of matter falls, as the universe expands (as the distance between galaxies becomes ever greater and matter there-fore gets "diluted"), just as you would expect. However, the energy density in empty space remains constant, because, one might argue, with empty space there is nothing to dilute! (Or, as I have somewhat less facetiously described, the universe does work on empty space as it expands.) The two curves cross rela-tively close to the present time, which is the source of the strange coincidence I have described.

Now consider what would happen if the energy in empty space were, say, 50 times greater than the value we estimate today. Then the two curves would cross at a different, earlier time, as shown in the figure below.

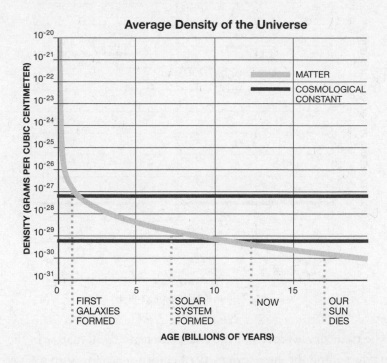

The time that the two curves cross for the upper, enlarged value of the energy of empty space is the time when galaxies first formed, about a billion years after the Big Bang. But remember that the energy of empty space is gravitationally repulsive. If it had come to dominate the energy of the universe before the time of galaxy formation, the repulsive force due to this energy would have outweighed (literally) the normal attractive gravitational force that caused matter to clump together. And galaxies would never have formed!

But if galaxies hadn't formed, then stars wouldn't have formed. And if stars hadn't formed, planets wouldn't have formed. And if planets hadn't formed, then astronomers wouldn't have formed!

So, in a universe with an energy of empty space merely 50 times bigger than that we observe, apparently no one would have been around today to try to measure the energy.

Could this be telling us something? Shortly after the discovery of our accelerating universe, physicist Steven Weinberg proposed, based on an argument he had developed more than a decade earlier—before the discovery of dark energy—that the "Coincidence Problem" could therefore be solved if perhaps the value of the cosmological constant that we measure today were somehow "anthropically" selected. That is, if somehow there were many universes, and in each universe the value of the energy of empty space took a randomly chosen value based on some probability distribution among all possible energies, then only in those universes in which the value is not that different from what we measure would life as we know it be able to evolve. So maybe we find ourselves in a universe with a tiny energy in empty space because we couldn't find ourselves in one with a much larger value. Put another way, it is not too surprising to find that we live in a universe in which we can live!

This argument, however, makes mathematical sense only if there is a possibility that many different universes have arisen. Talking about many different universes can sound like an oxymoron. After all, traditionally the notion of universe has become synonymous with "everything that exists."

More recently, however, *universe* has come to have a simpler, arguably more sensible meaning. It is now traditional to think of "our" universe as comprising simply the totality of all that we can now see and all that we could ever see. Physically, therefore, our universe comprises everything that either once could have had an impact upon us or that ever will.

The minute one chooses this definition for a universe, the possibility of other "universes"—regions that have always been and always will be causally disconnected from ours, like islands separated from any communication with one another by an ocean of space—becomes possible, at least in principle.

Our universe is so vast that, as I have emphasized, something that is not impossible is virtually guaranteed to occur somewhere within it. Rare events happen all the time. You might wonder whether the same principle applies to the possibility of many universes, or a *multiverse,* as the idea is now known. It turns out that the theoretical situation is actually stronger than simply a possibility. A number of central ideas that drive much of the current activity in particle theory today appear to require a multiverse.

I want to stress this because, in discussions with those who feel the need for a creator, the existence of a multiverse is viewed as a cop-out conceived by physicists who have run out of answers—or perhaps questions. This may eventually be the case, but it is not so now. Almost every logical possibility we can imagine regarding extending laws of physics as we know them, on small scales, into a more complete theory, suggests that, on large scales, our universe is not unique.

The phenomenon of inflation provides perhaps the first, and perhaps best, rationale. In the inflationary picture, during the phase when a huge energy temporarily dominates some region of the universe, this region begins to expand exponentially. At some point, a small region within this "false vacuum" may exit inflation as a phase transition occurs within the region and the field within it relaxes to its true, lower energy value; the expansion within this region will then cease to be exponential. But the space *between* such regions will continue to expand exponentially. At any one time, unless the phase transition completes through all of space, then almost all of space lies within an inflating region. And the inflating region will separate those regions that first exit inflation by almost unfathomable distances. It is like lava pouring out of a volcano. Some of the rock will cool and solidify, but those rocks will be carried far apart from one another as they float on a sea of liquid magma.

The situation can be even more dramatic. In 1986, Andrei Linde, who along with Alan Guth has been one of the chief architects of modern inflationary theory, promoted and explored a possibly even more general scenario. This was also anticipated in some sense by another inventive Russian cosmologist in the United States, Alex Vilenkin. Both Linde and Vilenkin have the inner confidence that one finds in great Russian physicists, but their history is quite different. Linde thrived in the old Soviet physics establishment before immigrating to the United States after the fall of the Soviet Union. Brash, brilliant, and funny, he has continued to dominate much of theoretical particle cosmology in the interim. Vilenkin emigrated far earlier, before he was a physicist, and worked in the then Soviet Union in various jobs, including as a night watchman, before he moved, as he was blacklisted by the KGB and could not get into graduate school there. And while he was always interested in cosmology, he accidentally applied to the wrong school for graduate work and ended up doing a thesis in condensed matter physics—the physics of materials. He then got a job as a postdoctoral researcher at Case Western Reserve University, where I later became Chair. During this period, he asked his supervisor, Philip Taylor, if he could spend several days a week working on cosmology in addition to his assigned projects. Philip later told me that, even with this part-time labor, Alex was the most productive postdoc he had ever had.

In any case, what Linde recognized is that, while quantum fluctuations during inflation may often push the field that drives inflation toward its lowest energy state, and thus provide a graceful exit, there is always the possibility that, in some regions, quantum fluctuations will drive the field toward yet higher energies, and hence away from values where inflation will end, so that inflation will continue unabated. Because such regions will expand for longer periods of time, there will be far more space that is inflating than that which is not. Within these regions, quantum fluctuations again will drive some subregions to exit inflation and thus stop expanding exponentially, but again there will be regions where quantum fluctuations will cause inflation to persist even longer. And so on.

This picture, which Linde dubbed "chaotic inflation," indeed resembles more familiar chaotic systems on Earth. Take boiling oatmeal, for example. At any point a bubble of gas may burst from the surface, reflecting regions where liquid at high temperature completes a phase transition to form a vapor. But between the bubbles the oatmeal is roiling and flowing. On large scales there is regularity—there are always bubbles popping somewhere. But locally, things are quite different depending upon where one looks. So it would be in a chaotically inflating universe. If one happened to be located in a "bubble" of true ground state that had stopped inflating, one's universe would appear very different from the vast bulk of space around it, which would still be inflating.

In this picture, inflation is eternal. Some regions, indeed most of space, will go on inflating forever. Those regions that exit inflation will become separate, causally disconnected universes. I want to stress that a multiverse is *inevitable* if inflation is eternal, and eternal inflation is by far the most likely possibility in most, if not all, inflationary scenarios. As Linde put it in his 1986 paper:

> The old question why our universe is the only possible one is now replaced by the question in which theories [of] the existence of mini-universes of our type [are] possible. This question is still very difficult, but it is much easier than the previous one. In our opinion, the modification of the point of view on the global structure of the universe and on our place in the world is one of the most important consequences of the development of the inflationary universe scenario.

As Linde emphasized, and has since become clear, this picture also provides another new possibility for physics. It could easily be that there are many possible low-energy quantum states of the universe present in nature that an inflating universe might ultimately decay into. Because the configuration of the quantum states of these fields will be different in each such region, the character of the fundamental laws of physics in each region/universe can then appear different.

Here arose the first "landscape" in which the anthropic argument, provided earlier, could play itself out. If there are many different states in which our universe could end up in after inflation, perhaps the one we live in, one in which there is non-zero vacuum energy that is small enough so galaxies could form, is just one of a potentially infinite family and the one that is selected for inquisitive scientists because it supports galaxies, stars, planets, and life.

The term "landscape" did not, however, first arise in this context. It was promoted by a much more effective marketing machine associated with the juggernaut that has been driving particle theory for much of the past quarter century—string theory. String theory posits that elementary particles are made up of more fundamental constituents, not particles, but objects that behave like vibrating strings. Just as string vibrations on a violin can create different notes, so too in this theory different sorts of vibrations produce objects that might, in principle, behave like all the different elementary particles we find in nature. The catch, however, is that the theory is not mathematically consistent when defined in merely four dimensions, but appears to require many more to make sense. What happens to the other dimensions is not immediately obvious, nor is the issue of what other objects besides strings may be important to define the theory—just some of the many unsolved challenges that have presented themselves and dulled some of the early enthusiasm for this idea.

Here is not the place to thoroughly review string theory, and in fact a thorough review is probably not possible, because if one thing has become clear in the past twenty-five years, it is that what was formerly called string theory is clearly something much more elaborate and complicated, and something whose fundamental nature and makeup is still a mystery.

We still have no idea if this remarkable theoretical edifice actually has anything to do with the real world. Nevertheless, perhaps no theoretical picture has ever so successfully permeated the consciousness of the physics community without having yet demonstrated its ability to successfully resolve a single experimental mystery about nature.

Many people will take the last sentence as a criticism of string theory, but although I have been branded in the past as a detractor, that is not really my intent here, nor has it been my intent in the numerous lectures and well-intentioned public debates I have had with my friend Brian Greene, one of string theory's main proponents, on the subject. Rather, I think it is simply important to cut through the popular hype for a reality check. String theory involves fascinating ideas and mathematics that might shed light on one of the most fundamental inconsistencies in theoretical physics—our inability to cast Einstein's general relativity in a form that can be combined with the laws of quantum mechanics to result in sensible predictions about how the universe behaves on its very smallest scales.

I have written a whole book about how string theory has attempted to circumvent this problem, but for our purposes here, only a very brief summary is necessary. The central proposal is simple to state, if difficult to implement. On very small scales, appropriate to the scale where the problems between gravity and quantum mechanics might first be encountered, elementary strings may curl up into closed loops. Amidst the set of excitations of such closed loops there always exists one such excitation that has the properties of the particle that, in quantum theory, conveys the force of gravity—the graviton. Thus, the quantum theory of such strings provides, in principle, the playing field on which a true quantum theory of gravity might be built.

Sure enough, it was discovered that such a theory might avoid the embarrassing infinite predictions of the standard quantum approaches to gravity. There was one hitch, however. In the simplest version of the theory, such infinite predictions can be obviated only if the strings that make up elementary particles are vibrating, not merely in the three dimensions of space and one of time that we are all familiar with, but rather in twenty-six dimensions!

You might expect that such a leap of complexity (and, perhaps, faith) would be enough to turn off most physicists about the theory, but in the mid-1980s some beautiful mathematical work by a host of individuals, most notably Edward Witten at the Institute for Advanced Study, demonstrated that the theory could in principle do far more than just provide a quantum theory of gravity. By introducing new mathematical symmetries, most notably a remarkably powerful mathematical framework called "supersymmetry," it became possible to reduce the number of dimensions required for consistency of the theory from twenty-six to merely ten.

More important, however, it looked like it might be possible, within the context of string theory, to unify gravity with the other forces in nature in a single theory, and moreover possible to explain the existence of every single elementary particle known in nature! Finally, it appeared as if there might be a single unique theory in ten dimensions that would reproduce everything we see in our four-dimensional world.

Claims of a "Theory of Everything" began to propagate, not just in the scientific literature, but in popular literature as well. As a result, perhaps more people are familiar with "superstrings" than are familiar with "superconductivity"—the latter being the remarkable fact that when some materials are cooled to extremely low temperatures, they can conduct electricity without any resistance whatsoever. This is not only one of the most remarkable properties of matter ever observed, but it has already transformed our understanding of the quantum makeup of materials.

Alas, the intervening twenty-five years or so have not been kind to string theory. Even as the best theoretical minds in the world began to focus their attention on it, producing volumes of new results and a great deal of new mathematics in the process (Witten went on to win the highest prize in mathematics, for example), it became clear that the "strings" in string theory are probably not the fundamental objects at all. Other, more complicated structures, called "branes," named after membranes in cells, which exist in higher dimensions, probably control the behavior of the theory.

What is worse, the uniqueness of the theory began to disappear. After all, the world of our experience is not ten-dimensional, but rather four-dimensional. Something has to happen to the remaining six spatial dimensions, and the canonical explanation of their invisibility is that they are somehow "compactified"—that is, they are curled up on such small scales that we cannot resolve them on our scales or even on the tiny scales that are probed by our highest energy particle accelerators today.

There is a difference between these proposed hidden domains and the domains of spirituality and religion, even though they may not appear so different on the surface. In the first place, they are accessible in principle if one could build a sufficiently energetic accelerator—beyond the bounds of practicality perhaps, but not beyond the bounds of possibility. Second, one might hope, as one does for virtual particles, to find some indirect evidence of their existence via the objects we can measure in our four-dimensional universe. In short, because these dimensions were proposed as part of a theory developed to actually attempt to explain the universe, rather than justify it, they might ultimately be accessible to empirical testing, even if the likelihood is small.

But beyond this, the possible existence of these extra dimensions provides a huge challenge to the hope that our universe is unique. Even if one starts with a unique theory in ten dimensions (which, I repeat, we do not yet know exist), then every different way of compactifying the invisible six dimensions can result in a different type of four-dimensional universe, with different laws of physics, different forces, different particles, and governed by differing symmetries. Some theorists have esti-mated that there are perhaps 10^{500} different possible consistent four-dimensional universes that could result from a single ten-dimensional string theory. A "Theory of Everything" had sud-denly become a "Theory of Anything"!

This situation was exemplified sarcastically in a cartoon from one of my favorite scientific comic strips, called *xkcd*. In this strip one person says to another: "I just had an awesome idea. What if all matter and energy is made of tiny vibrating strings." The second person then says, "Okay. What would that imply?" To which the first person responds: "I dunno."

On a slightly less facetious note, the Nobel Prize–winning physicist Frank Wilczek has suggested that string theorists have invented a new way of doing physics, reminiscent of a novel way of playing darts. First, one throws the dart against a blank wall, and then one goes to the wall and draws a bull's-eye around where the dart landed.

While Frank's comment is an accurate reflection of much of the hype that has been generated, it should be stressed that at the same time those working on the theory are honestly trying to uncover principles that might govern the world in which we live. Nevertheless, the plethora of possible four-dimensional uni-verses, which used to be such an embarrassment for string theo-rists, has now become a virtue of the theory. One can imagine that, in a ten-dimensional "multiverse" one can embed a host of different four-dimensional universes (or five-dimensional ones, or six-dimensional ones, or so on . . .), and each one can have dif-ferent laws of physics, and moreover, in each one the energy of empty space can be different.

While it sounds like a convenient fabrication, it appears to be an automatic consequence of the theory, and it does create a true multiverse "landscape" that might provide a natural framework for developing an anthropic understanding of the energy of empty space. In this case, we do not need an infinite number of possible universes separated in three-dimensional space. Rather, we can imagine an infinite number of universes stacked up above a single point in our space, invisible to us, but each of which could exhibit remarkably different properties.

I want to emphasize that this theory is not as trivial as the theological musing of Saint Thomas Aquinas about whether several angels could occupy the same place, an idea that was derided by later theologians as fruitless speculations on how many angels could fit on the point of a needle — or most popularly, on the head of a pin. Aquinas actually answered this question himself by saying that more than one angel could not occupy the same space — of course, without any theoretical or experimental justification! (And if they were bosonic quantum angels, he would have been wrong in any case.)

Presented with such a picture, and adequate mathematics, one might hope, in principle, to actually make physical predictions. For example, one might derive a probability distribution describing the likelihood of finding different types of four-dimensional universes embedded in a larger dimensional multiverse. One might find, for example, that the bulk of such universes that have small vacuum energy also have three families of elementary particles and four different forces. Or one might find that only in universes with small vacuum energy could there exist a long-range force of electromagnetism. Any such result might provide reasonably compelling evidence that a probabilistic anthropic explanation of the energy of empty space — in other words, finding that a universe that looks like ours with small vacuum energy is not improbable — makes solid physical sense.

Yet the mathematics has not yet brought us this far, and it may never do so. But in spite of our current theoretical impotence, this does not mean that this possibility is not actually realized by nature.

Nevertheless, in the meantime, particle physics has taken anthropic reasoning a step further.

Particle physicists are way ahead of cosmologists. Cosmology has produced one totally mysterious quantity: the energy of empty space, about which we understand virtually nothing. However, particle physics has not understood many more quantities for far longer!

For example: Why are there three generations of elementary particles—the electron, and its heavier cousins the muon and tauon, for example, or the three different sets of quarks, of which the lowest energy set makes up the bulk of matter we find on Earth? Why is gravity so much weaker than the other forces in nature, such as electromagnetism? Why is the proton 2,000 times heavier than the electron?

Some particle physicists have now jumped on the anthropic bandwagon in the extreme, perhaps because their efforts to explain these mysteries according to physical causes have not yet been successful. After all, if one fundamental quantity in nature is actually an environmental accident, why aren't most or all of the other fundamental parameters? Maybe all of the mysteries of particle theory can be solved by invoking the same mantra: if the universe were any other way, we could not live in it.

One might wonder if such a solution of the mysteries of nature is any solution at all or, more important, whether it describes science as we understand it. After all, the goal of science, and in particular physics, over the past 450 years has been to explain why the universe must be the way we measure it to be, rather than why in general the laws of nature would produce universes that are quite different.

I have tried to explain why this is not quite the case, namely why many respectable scientists have turned to the anthropic principle and why a number have worked quite hard to see if we might learn something new about our universe based on it.

Let me now go further and try to explain how the existence of forever undetectable universes—either removed from us by virtually infinite distances in space or, right beyond the tip of our noses, removed from us by microscopic distances in possible extra dimensions—might nevertheless be subject to some kind of empirical testing.

Imagine, for example, that we devised a theory based on unifying at least three of the four forces of nature in some Grand Unified Theory, a subject of continued intense interest in particle physics (among those who have not given up looking for fundamental theories in four dimensions). Such a theory would make predictions about the forces of nature that we measure and about the spectrum of elementary particles that we probe at our accelerators. Should such a theory make a host of predictions that are subsequently verified in our experiments, we would have very good reason to suspect that it contains a germ of truth.

Now, suppose this theory also predicts a period of inflation in the early universe, and in fact predicts that our inflationary epoch is merely one of a host of such episodes in an eternally inflating multiverse. Even if we could not explore the existences of such regions beyond our horizon directly, then, as I have said earlier, if it walks like a duck and quacks like a duck . . . Well, you know.

Finding possible empirical support for the ideas surrounding extra dimensions is more far-fetched but not impossible. Many bright young theorists are devoting their professional careers to the hope of developing the theory to the point where there might be some evidence, even indirect, that it is correct. Their hopes might be misplaced, but they have voted with their feet.

Perhaps some evidence from the new Large Hadron Collider near Geneva will reveal some otherwise hidden window into this new physics.

So, after a century of remarkable, truly unprecedented progress in our understanding of nature, we have found ourselves able to probe the universe on scales that were previously unimaginable. We have understood the nature of the Big Bang expansion back to its earliest microseconds and have discovered the existence of hundreds of billions of new galaxies, with hundreds of billions of new stars. We have discovered that 99 percent of the universe is actually invisible to us, comprising dark matter that is most likely some new form of elementary particle, and even more dark energy, whose origin remains a complete mystery at the present time.

And after all of this, it may be that physics will become an "environmental science." The fundamental constants of nature, so long assumed to take on special importance, may just be environmental accidents. If we scientists tend to take ourselves and our science too seriously, maybe we also have taken our universe too seriously. Maybe literally, as well as metaphorically, we are making much ado about nothing. At least we may be making too much of the nothing that dominates our universe! Maybe our universe is rather like a tear buried in a vast multiversal ocean of possibilities. Maybe we will never find a theory that describes why the universe has to be the way it is.

Or maybe we will.

That, finally, is the most accurate picture I can paint of reality as we now understand it. It is based on the work of tens of thousands of dedicated minds over the past century, building some of the most complex machines ever devised and developing some of the most beautiful and also the most complex ideas with which humanity has ever had to grapple. It is a picture whose creation emphasizes the best about what it is to be human—our ability to imagine the vast possibilities of existence and the adventurousness to bravely explore them—without passing the buck to a vague creative force or to a creator who is, by definition, forever unfathomable. We owe it to ourselves to draw wisdom from this experience. To do otherwise would do a disservice to all the brilliant and brave individuals who helped us reach our current state of knowledge.

If we wish to draw philosophical conclusions about our own existence, our significance, and the significance of the universe itself, our conclusions should be based on empirical knowledge. A truly open mind means forcing our imaginations to conform to the evidence of reality, and not vice versa, whether or not we like the implications.

CHAPTER 9

NOTHING IS SOMETHING

I don't mind not knowing. It doesn't scare me.
—RICHARD FEYNMAN

Isaac Newton, perhaps the greatest physicist of all time, profoundly changed the way we think about the universe in many ways. But perhaps the most important contribution he made was to demonstrate the possibility that the entire universe is explicable. With his universal law of gravity, he demonstrated for the first time that even the heavens might bend to the power of natural laws. A strange, hostile, menacing, and seemingly capricious universe might be nothing of the sort.

If immutable laws governed the universe, the mythical gods of ancient Greece and Rome would have been impotent. There would have been no freedom to arbitrarily bend the world to create thorny problems for mankind. What held for Zeus would also apply to the God of Israel. How could the Sun stand still at midday if the Sun did not orbit the Earth but its motion in the sky was actually caused by the revolution of the Earth, which, if suddenly stopped, would produce forces on its surface that would destroy all human structures and humans along with them?

Of course, supernatural acts are what miracles are all about. They are, after all, precisely those things that circumvent the laws of nature. A god who can create the laws of nature can presumably also circumvent them at will. Although why they would have been circumvented so liberally thousands of years ago, before the invention of modern communication instruments that could have recorded them, and not today, is still something to wonder about.

In any case, even in a universe with no miracles, when you are faced with a profoundly simple underlying order, you can draw two different conclusions. One, drawn by Newton himself, and earlier espoused by Galileo and a host of other scientists over the years, was that such order was created by a divine intelligence responsible not only for the universe, but also for our own existence, and that we human beings were created in her image (and apparently other complex and beautiful beings were not!). The other conclusion is that the laws themselves are all that exist. These laws themselves require our universe to come into existence, to develop and evolve, and we are an irrevocable by-product of these laws. The laws may be eternal, or they too may have come into existence, again by some yet unknown but possibly purely physical process.

Philosophers, theologians, and sometimes scientists continue to debate these possibilities. We do not know for certain which of them actually describes our universe, and perhaps we shall never know. But the point is, as I emphasized at the very beginning of this book, the final arbiter of this question will not come from hope, desire, revelation, or pure thought. It will come, if it ever does, from an exploration of nature. Dream or nightmare, as Jacob Bronowski said in the opening quote in the book—and one person's dream in this case can easily be another's nightmare—we need to live our experience as it is and with our eyes open. The universe is the way it is, whether we like it or not.

And here, I think it is *extremely significant* that a universe from nothing—in a sense I will take pains to describe—that arises naturally, and even inevitably, is increasingly consistent with everything we have learned about the world. This learning has *not* come from philosophical or theological musings about morality or other speculations about the human condition. It is instead based on the remarkable and exciting developments in empirical cosmology and particle physics that I have described.

I want thus to return to the question I described at the beginning of this book: Why is there something rather than nothing? We are now presumably in a better position to address this, having reviewed the modern scientific picture of the universe, its history, and its possible future, as well as operational descriptions of what "nothing" might actually comprise. As I also alluded to at the beginning of this book, this question too has been informed by science, like essentially all such philosophical questions. Far from providing a framework that forces upon us the requirement of a creator, the very meaning of the words involved have so changed that the sentence has lost much of its original meaning—something that again is not uncommon, as empirical knowledge shines a new light on otherwise dark corners of our imagination.

At the same time, in science we have to be particularly cautious about "why" questions. When we ask, "Why?" we usually mean "How?" If we can answer the latter, that generally suffices for our purposes. For example, we might ask: "Why is the Earth 93 million miles from the Sun?" but what we really probably mean is, "How is the Earth 93 million miles from the Sun?" That is, we are interested in what physical processes led to the Earth ending up in its present position. "Why" implicitly suggests purpose, and when we try to understand the solar system in scientific terms, we do not generally ascribe purpose to it.

So I am going to assume what this question really means to ask is, "How is there something rather than nothing?" "How" questions are really the only ones we can provide definitive answers to by studying nature, but because this sentence sounds much stranger to the ear, I hope you will forgive me if I sometimes fall into the trap of appearing to discuss the more standard formulation when I am really trying to respond to the more specific "how" question.

Even here, from the perspective of actual *understanding,* this particular "how" question has been supplanted by a host of operationally more fruitful questions, such as, "What might have produced the properties of the universe that most strikingly characterize it at the present time?" or, perhaps more important, "How can we find out?"

Here I want to once again beat what I wish were a dead horse. Framing questions in this way allows the production of new knowledge and understanding. This is what differentiates them from purely theological questions, which generally presume the answers up front. Indeed, I have challenged several theologians to provide evidence contradicting the premise that theology has made no contribution to knowledge in the past five hundred years at least, since the dawn of science. So far no one has provided a counterexample. The most I have ever gotten back was the query, "What do you mean by knowledge?" From an epistemological perspective this may be a thorny issue, but I maintain that, if there were a better alternative, someone would have presented it. Had I presented the same challenge to biologists, or psychologists, or historians, or astronomers, none of them would have been so flummoxed.

The answers to these sorts of fruitful questions involve theoretical predictions that can be tested via experiments to drive our operational knowledge of the universe forward more directly.

Partly for this reason, I have focused on such fruitful questions up to this point in this book. Nevertheless, the "something from nothing" question continues to have great currency, and therefore probably needs to be confronted.

Newton's work dramatically reduced the possible domain of God's actions, whether or not you attribute any inherent rationality to the universe. Not only did Newton's laws severely constrain the freedom of action of a deity, they dispensed with various requirements for supernatural intervention. Newton discovered that the motion of planets around the Sun does not require them to be continually pushed along their paths, but rather, and highly nonintuitively, requires them to be pulled by a force acting toward the Sun, thus dispensing of the need for the angels who were often previously invoked as guiding the planets on their way. While dispensing with this particular use of angels has had little impact on people's willingness to believe in them (polls suggest far more people believe in angels in the United States than believe in evolution), it is fair to say that progress in science since Newton has even more severely constrained the available opportunities for the hand of God to be manifest in his implied handiwork.

We can describe the evolution of the universe back to the earliest moments of the Big Bang without specific need for anything beyond known physical laws, and we have also described the universe's likely future history. There are certainly still puzzles about the universe that we don't understand, but I am going to assume that readers of this book are not wedded to a "God of the Gaps" picture, whereby God is invoked whenever there is something specific about our observations that seems puzzling or not fully understood. Even theologians recognize that such recourse not only diminishes the grandeur of their supreme being, but it also opens that being up to being removed or further marginalized whenever new work explains or removes the puzzle.

In this sense, the "something from nothing" argument really tries to focus on the original act of creation and asks whether a scientific explanation can ever be logically complete and fully satisfying in addressing this specific issue.

It turns out that, given our current understanding of nature, there are three different, separate meanings for the "something from nothing" question. The short answer to each is "quite plausibly yes," and I shall discuss each in turn in the rest of this book as I attempt to explain why or, as I have argued just now, better yet how.

Occam's razor suggests that, if some event is physically plausible, we don't need recourse to more extraordinary claims for its being. Surely the requirement of an all-powerful deity who somehow exists outside of our universe, or multiverse, while at the same time governing what goes on inside it, is one such claim. It should thus be a claim of last, rather than first, resort.

I have already argued in the preface to this book that merely defining "nothingness" as "nonbeing" is not sufficient to suggest that physics, and more generally science, is not adequate to address the question. Let me give an additional, more specific argument here. Consider an electron-positron pair that spontaneously pops out of empty space near the nucleus of an atom and affects the property of that atom for the short time the pair exists. In what sense did the electron or positron exist before? Surely by any sensible definition they didn't. There was potential for their existence, certainly, but that doesn't define *being* any more than a potential human being exists because I carry sperm in my testicles near a woman who is ovulating, and she and I might mate. Indeed, the best answer I have ever heard to the question of what it would be like to be dead (i.e., be nonbeing) is to imagine how it felt to be before you were conceived. In any case, if potential to exist were the same as existence, then I am certain that by now masturbation would be as hot button a legal issue as abortion now is.

The Origins Project at Arizona State University, which I direct, recently ran a workshop on the Origin of Life, and I cannot help but view the present cosmological debate in this context. We do not yet fully understand how life originated on Earth. However, we have not only plausible chemical mechanisms by which this might be conceivable, but we are also homing in closer and closer every day to specific pathways that might have allowed biomolecules, including RNA, to arise naturally. Moreover, Darwinian evolution, based on natural selection, provides a compellingly accurate picture of how complex life emerged on this planet following whatever specific chemistry produced the first faithfully self-replicating cells with a metabolism that captured energy from their environment. (As good a definition of life as I can come up with for the moment.)

Just as Darwin, albeit reluctantly, removed the need for divine intervention in the evolution of the modern world, teeming with diverse life throughout the planet (though he left the door open to the possibility that God helped breathe life into the first forms), our current understanding of the universe, its past, and its future make it more plausible that "something" can arise out of nothing without the need for any divine guidance. Because of the observational and related theoretical difficulties associated with working out the details, I expect we may never achieve more than plausibility in this regard. But plausibility itself, in my view, is a tremendous step forward as we continue to marshal the courage to live meaningful lives in a universe that likely came into existence, and may fade out of existence, without purpose, and certainly without us at its center.

Let's now return to one of the most remarkable features of our universe: it is as close to being flat as we can measure. I remind you of the unique facet of a flat universe, at least on scales where it is dominated by matter in the form of galaxies, and where a Newtonian approximation remains valid: in a flat universe, and only in a flat universe, the average Newtonian gravitational energy of every object participating in the expansion is precisely zero.

I emphasize that this was a falsifiable postulate. It didn't have to be this way. Nothing required this except theoretical speculations based on considerations of a universe that could have arisen naturally from nothing, or at the very least, from *almost nothing*.

I cannot overstress the importance of the fact that, once gravity is included in our considerations of nature, one is no longer free to define the total energy of a system arbitrarily, nor the fact that there are both positive and negative contributions to this energy. Determining the total gravitational energy of objects being carried along by the expansion of the universe is *not* subject to arbitrary definition any more than the geometric curvature of the universe is a matter of definition. It is a property of space itself, according to general relativity, and this property of space is determined by the energy contained within it.

I say this because it has been argued that the statement that the average total Newtonian gravitational energy of every galaxy in a flat, expanding universe is zero is arbitrary, and that any other value would be just as good, but that scientists "define" the zero point to argue against God. So claimed Dinesh D'Souza, anyway, in his debates with Christopher Hitchens on the existence of God.

Nothing could be further from the truth. The effort to determine the curvature of the universe was an undertaking carried out over half a century by scientists who devoted their lives to determining the actual nature of the universe, not to imposing their own desires upon it. Even well after the theoretical arguments about why the universe should be flat were first proposed, my observational colleagues, during the 1980s and even early 1990s, remained bent on proving otherwise. For, after all, in science one achieves the greatest impact (and often the greatest headlines) not by going along with the herd, but by bucking against it.

Nevertheless, the data have had the last word, and the last word is in. Our observable universe is as close to being flat as we can measure. The Newtonian gravitational energy of galaxies moving along with the Hubble expansion *is* zero—like it or not.

I would now like to describe how, if our universe arose from nothing, a flat universe, one with zero total Newtonian gravitational energy of every object, is precisely what we should expect. The argument is a little subtle—subtler than I have been able to describe in my popular lectures on the subject—so I am happy to have the space here to carefully try to lay it out.

First, I want to be clear about what kind of "nothing" I am discussing at the moment. This is the simplest version of nothing, namely empty space. For the moment, I will assume space exists, with nothing at all in it, and that the laws of physics also exist. Once again, I realize that in the revised versions of nothingness that those who wish to continually redefine the word so that no scientific definition is practical, this version of nothing doesn't cut the mustard. However, I suspect that, at the times of Plato and Aquinas, when they pondered why there was something rather than nothing, empty space with nothing in it was probably a good approximation of what they were thinking about.

As we saw in chapter 6, Alan Guth has explained precisely how we can get something from this kind of nothing—the ultimate free lunch. Empty space can have a non-zero energy associated with it, even in the absence of any matter or radiation. General relativity tells us that space will expand exponentially, so that even the tiniest region at early times could quickly encompass a size more than large enough to contain our whole visible universe today.

As I also described in that chapter, during such a rapid expansion, the region that will eventually encompass our universe will get flatter and flatter even as the energy contained within empty space grows as the universe grows. This phenomenon happens without the need for any hocus pocus or miraculous intervention. This is possible because the gravitational "pressure" associated with such energy in empty space is actually negative. This "negative pressure" implies that, as the universe expands, the expansion dumps energy *into* space rather than vice versa.

According to this picture, when inflation ends, the energy stored in empty space gets turned into an energy of real particles and radiation, creating effectively the traceable beginning of our present Big Bang expansion. I say the traceable beginning because inflation effectively erases any memory of the state of the universe before it began. All complexities and irregularities on initially large scales (if the initial preexisting universe or metaverse were large, even infinitely large) get smoothed out and/or driven so far outside our horizon today that we will always observe an almost uniform universe after enough inflationary expansion has taken place.

I say almost uniform because I also described in chapter 6 how quantum mechanics will always leave some residual, small-density fluctuations that get frozen during inflation. This results in the second amazing implication of inflation, that small-density fluctuations in empty space due to the rules of quantum mechanics will later be responsible for all the structure we observe in the universe today. So we, and everything we see, result out of quantum fluctuations in what is essentially nothingness near the beginning of time, namely during the inflationary expansion.

After all the dust is settled, the generic configuration of the matter and radiation will be that of an essentially flat universe, one in which the average Newtonian gravitational energy of all objects will appear to be zero. This will almost always be the case, unless one could very carefully fine-tune the amount of inflation.

Therefore, our observable universe can start out as a microscopically small region of space, which can be essentially empty, and still grow to enormous scales containing eventually lots of matter and radiation, all without costing a drop of energy, with enough matter and radiation to account for everything we see today!

The important point worth stressing in this brief summary of the inflationary dynamics discussed in chapter 6 is that something can arise from empty space *precisely* because the energetics of empty space, in the presence of gravity, are *not* what common sense would have guided us to suspect before we discovered the underlying laws of nature.

But no one ever said that the universe is guided by what we, in our petty myopic corners of space and time, might have originally thought was sensible. It certainly seems sensible to imagine that a priori, matter cannot spontaneously arise from empty space, so that *something*, in this sense, cannot arise from *nothing*. But when we allow for the dynamics of gravity and quantum mechanics, we find that this commonsense notion is no longer true. This is the *beauty* of science, and it should not be threatening. Science simply forces us to revise what is sensible to accommodate the universe, rather than vice versa.

To summarize then: the observation that the universe is flat and that the local Newtonian gravitational energy is essentially zero today is strongly suggestive that our universe arose through a process like that of inflation, a process whereby the energy of empty space (nothing) gets converted into the energy of something, during a time when the universe is driven closer and closer to being essentially exactly flat on all observable scales.

While inflation demonstrates how empty space endowed with energy can effectively create everything we see, along with an unbelievably large and flat universe, it would be disingenuous to suggest that empty space endowed with energy, which drives inflation, is really *nothing*. In this picture one must assume that space exists and can store energy, and one uses the laws of physics like general relativity to calculate the consequences. So if we stopped here, one might be justified in claiming that modern science is a long way from really addressing how to get something from nothing. This is just the first step, however. As we expand our understanding, we will next see that inflation can represent simply the tip of a cosmic iceberg of nothingness.

CHAPTER 10

Nothing Is Unstable

Fiat justitia—ruat caelum.
(Do justice, and let the skies fall.)
—ANCIENT ROMAN PROVERB

The existence of energy in empty space—the discovery that rocked our cosmological universe and the idea that forms the bedrock of inflation—only reinforces something about the quantum world that was already well established in the context of the kinds of laboratory experiments I have already described. Empty space is complicated. It is a boiling brew of virtual particles that pop in and out of existence in a time so short we cannot see them directly.

Virtual particles are manifestations of a basic property of quantum systems. At the heart of quantum mechanics is a rule that sometimes governs politicians or CEOs—as long as no one is watching, anything goes. Systems continue to move, if just momentarily, between all possible states, including states that would not be allowed if the system were actually being measured. These "quantum fluctuations" imply something essential about the quantum world: nothing always produces something, if only for an instant.

But here's the rub. The conservation of energy tells us that quantum systems can misbehave for only so long. Like embez-

153

zling stockbrokers, if the state that a system fluctuates into requires sneaking some energy from empty space, then the system has to return that energy in a time short enough so that no one measuring the system can detect it.

As a result, you might presume to safely argue that this "something" that is produced by quantum fluctuations is ephemeral—not measurable, unlike, say, you or I or the Earth on which we live. But this ephemeral creation, too, is subject to the circumstances associated with our measurements. For example, consider the electric field emanating from a charged object. It is definitely real. You can feel the static electric force on your hair or watch a balloon stick to a wall. However, the quantum theory of electromagnetism suggests that the static field is due to the emission, by the charged particles involved in producing the field, of virtual photons that have essentially zero total energy. These virtual particles, because they have zero energy, can propagate across the universe without disappearing, and the field due to the superposition of many of them is so real it can be felt.

Sometimes conditions are such that real, massive particles can actually pop out of empty space with impunity. In one example, two charged plates are brought close together and, once the electric field gets strong enough between them, it becomes energetically favorable for a real particle-antiparticle pair to "pop" out of the vacuum, with the negative charge heading toward the positive plate and the positive charge toward the negative one. In so doing, it is possible that the reduction in energy arising from reducing the net charge on each of the plates and hence the electric field between them can be greater than the energy associated with the rest mass energy required to produce two real particles. Of course, the strength of the field has to be huge for such a condition to be possible.

There is actually a place where strong fields of a different kind might allow a phenomenon similar to that described above to occur—but in this case due to gravity. This realization actually made Stephen Hawking famous among physicists in 1974, when he showed that it might be possible for black holes—out of which, in the absence of quantum mechanical considerations at least, nothing can ever escape—to radiate physical particles.

There are many different ways to try to understand this phenomenon, but one of these is strikingly familiar to the situation I described above with electric fields. Outside of the core of black holes is a radius called the "event horizon." Inside an event horizon, no object can classically escape because the escape velocity exceeds the speed of light. Thus, even light emitted inside this region will not make it outside the event horizon.

Now imagine a particle-antiparticle pair nucleates out of empty space just outside of the event horizon due to quantum fluctuations in that region. It is possible, if one of the particles actually falls within the event horizon, for it to lose enough gravitational energy by falling into the black hole that this energy exceeds twice the rest mass of either particle. This means that the partner particle can fly off to infinity and be observable without any violation of energy conservation. The total positive energy associated with the radiated particle is more than compensated by the loss of energy experienced by its partner particle falling into the black hole. The black hole can therefore radiate particles.

The situation is even more interesting, however, precisely because the energy lost by the infalling particle is greater than the positive energy associated with its rest mass. As a result, when it falls into the black hole, the net system of the black hole plus the particle actually has less energy than it did before the particle fell in! The black hole therefore actually gets *lighter* after the particle falls in by an amount that is equivalent to the energy carried away by the radiated particle that escapes. Eventually the black hole may radiate away entirely. At this point we do not know because the final stages of black hole evaporation involve physics on such small distance scales that general relativity alone cannot tell us the final answer. On these scales, gravity must be treated as a fully quantum mechanical theory, and our current understanding of general relativity is not sufficient to allow us to determine precisely what will happen.

Nevertheless, all of these phenomena imply that, under the right conditions, not only can nothing become something, it is required to.

An early example in cosmology of the fact that "nothing" can be unstable and form something comes from efforts to understand why we live in a universe of matter.

You probably don't wake up each morning wondering about this, but the fact that our universe contains matter is remarkable. What is particularly remarkable about this is that, as far as we can tell, our universe does not contain substantial amounts of antimatter, which you will recall is required to exist by quantum mechanics and relativity, so that for every particle that we know of in nature, there can exist an equivalent antiparticle with opposite charge and the same mass. Any sensible universe at its inception, one might think, would contain equal amounts of both. After all, the antiparticles of normal particles have the same mass and similar other properties, so if particles were created at early times, it would have been equally easy to create antiparticles.

Alternatively, we could even imagine an antimatter universe in which all of the particles that make up the stars and galaxies were replaced with their antiparticles. Such a universe would appear to be almost identical to the one we live in. Observers in such a universe (themselves made of antimatter) would no doubt call what we call antimatter as matter. The name is arbitrary.

However, if our universe began sensibly, with equal amounts

of matter and antimatter, and stayed that way, we wouldn't be around to ask "Why?" or "How?" This is because all particles of matter would have annihilated with all particles of antimatter in the early universe, leaving nothing but pure radiation. No matter or antimatter would be left over to make up stars, or galaxies, or to make up lovers or antilovers who might otherwise one day gaze out and be aroused by the spectacle of the night sky in each other's arms. No drama. History would consist of emptiness, a radiation bath that would slowly cool, leading ultimately to a cold, dark, bleak universe. Nothingness would reign supreme.

Scientists began to understand in the 1970s, however, that it is possible to begin with equal amounts of matter and antimatter in an early hot, dense Big Bang, and for plausible quantum processes to "create something from nothing" by establishing a small asymmetry, with a slight excess of matter over antimatter in the early universe. Then, instead of complete annihilation of matter and antimatter, leading to nothing but pure radiation today, all of the available antimatter in the early universe could have annihilated with matter, but the small excess of matter would have had no comparable amount of antimatter to annihilate with, and would then be left over. This would then lead to all the matter making up stars and galaxies we see in the universe today.

As a result, what might otherwise seem a small accomplishment (establishing a small asymmetry at early times) might instead be considered almost as the moment of creation. Because once an asymmetry between matter and antimatter was created, nothing could later put it asunder. The future history of a universe full of stars and galaxies was essentially written. Antimatter particles would annihilate with the matter particles in the early universe, and the remaining excess of matter particles would survive through the present day, establishing the character of the visible universe we know and love and inhabit.

Even if the asymmetry were 1 part in a billion there would be enough matter left over to account for everything we see in the universe today. In fact, an asymmetry of 1 part in a billion or so is precisely what was called for, because today there are roughly 1 billion photons in the cosmic microwave background for every proton in the universe. The CMBR photons are the remnants, in this picture, of the early matter-antimatter annihilations near the beginning of time.

A definitive description of how this process could have happened in the early universe is currently lacking because we have not yet fully and empirically established the detailed nature of the microphysical world at the scales where this asymmetry was likely to have been generated. Nevertheless, a host of different plausible scenarios has been explored based on the current best ideas we have about physics at these scales. While they differ in the details, they all have the same general characteristics. Quantum processes associated with elementary particles in the primordial heat bath can inexorably drive an empty universe (or equivalently an initially matter-antimatter symmetric universe) almost imperceptibly toward a universe that will be dominated by matter or antimatter.

If it could have gone either way, was it then just a circumstantial accident that our universe became dominated by matter? Imagine standing on top of a tall mountain and tripping. The direction you fall was not preordained, but rather is an accident, depending upon which direction you were looking in or at what point in your stride you tripped. Perhaps similarly our universe is like that, and even if the laws of physics are fixed, the ultimate direction of the asymmetry between matter and antimatter was driven by some random initial condition (just as in the case of tripping down the mountain, the law of gravity is fixed and determines that you will fall, but your direction may be an accident). Once again, our very existence in that case would be an environmental accident.

Independent of this uncertainty, however, is the remarkable fact that a feature of the underlying laws of physics can allow quantum processes to drive the universe away from a featureless state. Physicist Frank Wilczek, who was one of the first theorists to explore these possibilities, has reminded me that he utilized precisely the same language I have used previously in this chapter, in the 1980 *Scientific American* article he wrote on the matter-antimatter asymmetry of the universe. After describing how a matter-antimatter asymmetry might plausibly be generated in the early universe based on our new understanding of particle physics, he added a note that this provided one way of thinking about the answer to the question of why there is something rather than nothing: *nothing* is unstable.

The point Frank was emphasizing is that the measured excess of matter over antimatter in the universe appears on first glance to be an obstacle to imagining a universe that could arise from an instability in empty space, with nothingness producing a Big Bang. But if that asymmetry could arise dynamically after the Big Bang, that barrier is removed. As he put it:

> One can speculate that the universe began in the most symmetrical state possible and that in such a state no matter existed; the universe was a vacuum. A second state existed, and in it matter existed. The second state had slightly less symmetry, but was also lower in energy. Eventually a patch of less symmetrical phase appeared and grew rapidly. The energy released by the transition found form in the creation of particles. This event might be identified with the big bang . . . The answer to the ancient question "Why is there something rather than nothing?" would be that "nothing" is unstable.

Before I proceed, however, I am again reminded of the similarities between the discussion I have just given of a matter-antimatter asymmetry and the discussions we had at our recent Origins workshop to explore our current understanding of the nature of life in the universe and its origin. My words were different, but the fundamental issues are remarkably similar: What specific physical process in the early moments of the Earth's history could have led to the creation of the first replicating biomolecules and metabolism? As in the 1970s in physics, the recent decade has seen incredible progress in molecular biology. We learned of natural organic pathways, for example, that could produce, under plausible conditions, ribonucleic acids, long thought to be the precursors to our modern DNA-based world. Until recently it was felt that no such direct pathway was possible and that some other intermediate forms must have played a key role.

Now few biochemists and molecular biologists doubt that life can arise naturally from nonlife, even though the specifics are yet to be discovered. But, as we discussed all of this, a common subtext permeated our proceedings: Did the life that first formed on Earth *have* to have the chemistry that it did, or are there many different, equally viable possibilities?

Einstein once asked a question that, he said, was the one thing he really wanted to know about nature. I admit it is the most profound and fundamental question that many of us would like answered. He put it as follows: "What I want to know is whether *God* [*sic*] had any choice in the creation of the universe."

I have annotated this because Einstein's God was not the God of the Bible. For Einstein, the existence of order in the universe provided a sense of such profound wonder that he felt a spiritual attachment to it, which he labeled, motivated by Spinoza, with the moniker "God." In any case, what Einstein really meant in this question was the issue I have just described in the context of several different examples: Are the laws of nature unique? And is the universe we inhabit, which has resulted from these laws, unique? If you change one facet, one constant, one force, however slight, would the whole edifice crumble? In a biological sense, is the biology of life unique? Are we unique in the universe? We will return to discuss this most important question later in this book.

While such a discussion will cause us to further refine and generalize notions of "nothing" and "something," I want to return to taking an intermediate step in making the case for the inevitable creation of something.

As I have defined it thus far, the relevant "nothing" from which our observed "something" arises is "empty space." However, once we allow for the merging of quantum mechanics and general relativity, we can extend this argument to the case where space itself is forced into existence.

General relativity as a theory of gravity is, at its heart, a theory of space and time. As I described in the very beginning of this book, this means that it was the first theory that could address the dynamics not merely of objects moving through space, but also how space itself evolves.

Having a quantum theory of gravity would therefore mean that the rules of quantum mechanics would apply to the properties of space and not just to the properties of objects existing in space, as in conventional quantum mechanics.

Extending quantum mechanics to include such a possibility is tricky, but the formalism Richard Feynman developed, which led to a modern understanding of the origin of antiparticles, is well suited to the task. Feynman's methods focus on the key fact to which I alluded at the beginning of this chapter: quantum mechanical systems explore all possible trajectories, even those that are classically forbidden, as they evolve in time.

In order to explore this, Feynman developed a "sum over paths formalism" to make predictions. In this method, we consider all possible trajectories between two points that a particle might take. We then assign a probability weighting for each trajectory, based on well-defined principles of quantum mechanics, and then perform a sum over all paths in order to determine final (probabilistic) predictions for the motion of particles.

Stephen Hawking was one of the first scientists to fully exploit this idea to the possible quantum mechanics of space-time (the union of our three-dimensional space along with one dimension of time to form a four-dimensional unified space-time system, as required by Einstein's special theory of relativity). The virtue of Feynman's methods was that focusing on all possible paths ends up meaning that the results can be shown to be independent of the specific space and time labels one applies to each point on each path. Because relativity tells us that different observers in relative motion will measure distance and time differently and therefore assign different values to each point in space and time, having a formalism that is independent of the different labels that different observers might assign to each point in space and time is particularly useful.

And it is most useful perhaps in considerations of general relativity, where the specific labeling of space and time points becomes completely arbitrary, so that different observers at different points in a gravitational field measure distances and times differently, and all that ultimately determines the behavior of systems are geometric quantities like curvature, which turn out to be independent of all such labeling schemes.

As I have alluded to several times, general relativity is not fully consistent with quantum mechanics, at least as far as we can tell, and therefore there is no completely unambiguous method to define Feynman's sum-over-paths technique in general relativity. So we have to make some guesses in advance based on plausibility and check to see if the results make sense.

If we are to consider the quantum dynamics of space and time then, one must imagine that in the Feynman "sums," one must consider every different possible configuration that can describe the different geometries that space can adopt during the intermediate stages of any process, when quantum indeterminacy reigns supreme. This means we must consider spaces that are arbitrarily highly curved over short distances and small times (so short and so small that we cannot measure them so that quantum weirdness can reign free). These weird configurations would then not be observed by large classical observers such as us when we attempt to measure the properties of space over large distances and times.

But let's consider even stranger possibilities. Remember that, in the quantum theory of electromagnetism, particles can pop out of empty space at will as long as they disappear again on a time frame determined by the Uncertainty Principle. By analogy, then, in the Feynman quantum sum over possible space-time configurations, should one consider the possibility of small, possibly compact spaces that themselves pop in and out of existence? More generally, what about spaces that may have "holes" in them, or "handles" like donuts dunking into space-time?

These are open questions. However, unless one can come up with a good reason for excluding such configurations from the quantum mechanical sum that determines the properties of the evolving universe, and to date no such good reason exists that I know of, then under the general principle that holds everywhere else I know of in nature—namely that anything that is not proscribed by the laws of physics must actually happen—it seems most reasonable to consider these possibilities.

As Stephen Hawking has emphasized, a quantum theory of gravity allows for the creation, albeit perhaps momentarily, of space itself where none existed before. While in his scientific work he was not attempting to address the "something from nothing" conundrum, effectively this is what quantum gravity may ultimately address.

"Virtual" universes—namely the possible small compact spaces that may pop into and out of existence on a timescale so short we cannot measure them directly—are fascinating theoretical constructs, but they don't seem to explain how something can arise from nothing over the long term any more than do the virtual particles that populate otherwise empty space.

However, recall that a nonzero real electric field, observable at large distances away from a charged particle, can result from the coherent emission of many virtual zero energy photons by the charge. This is because virtual photons that carry zero energy do not violate energy conservation when they are emitted. The Heisenberg Uncertainty Principle, therefore, does not constrain them to exist for only very brief times before they must be reabsorbed and disappear back into nothingness. (Again recall that the Heisenberg Uncertainty Principle states that the uncertainty with which we measure the energy of a particle, and hence the possibility that its energy may change slightly by the emission and absorption of virtual particles, is inversely proportional to the length of time over which we observe it. Hence, virtual particles that carry away zero energy can do so essentially with impunity—namely they can exist for arbitrarily long times and travel arbitrarily far away before being absorbed . . . leading to the possible existence of long-range interactions between charged particles. If the photon was not massless, so that photons always carried away non-zero energy due to a rest mass, the Heisenberg Uncertainty Principle would imply that the electric field would be short range because photons could propagate only for short times without being reabsorbed again.)

A similar argument suggests that one can imagine one specific type of universe that might spontaneously appear and need not disappear almost immediately thereafter because of the constraints of the Uncertainty Principle and energy conservation. Namely, a compact universe with zero total energy.

Now, I would like nothing better than to suggest that this is precisely the universe we live in. This would be the easy way out, but I am more interested here in being true to our current understanding of the universe than in making an apparently easy and convincing case for creating it from nothing.

I have argued, I hope compellingly, that the average Newtonian gravitational energy of every object in our flat universe is zero. And it is. But that is not the whole story. Gravitational energy is not the total energy of any object. To this energy we must add its rest energy, associated with its rest mass. Put another way, as I have described earlier, the gravitational energy of an object at rest isolated from all other objects by an infinite distance is zero, because if it is at rest, it has no kinetic energy of motion, and if it is infinitely far away from all other particles, the gravitational force on it due to other particles, which could provide potential energy to do work, is also essentially zero. However, as Einstein told us, its total energy is not merely due to gravity, but also includes the energy associated with its mass, so that, as is famously known, $E = mc^2$.

In order to take this rest energy into account, we have to move from Newtonian gravity to general relativity, which, by definition, incorporates the effects of special relativity (and $E = mc^2$) into a theory of gravity. And here things get both subtler and more confusing. On small scales compared to the possible curvature of a universe, and as long as all objects within these scales are moving slowly compared to the speed of light, the general relativistic version of energy reverts to the definition we are familiar with from Newton. However, once these conditions no longer hold, all bets are off, almost.

Part of the problem is that it turns out that energy as we normally think of it elsewhere in physics is not a particularly well-defined concept on large scales in a curved universe. Different ways of defining coordinate systems to describe the different labels that different observers may assign to points in space and time (called different "frames of reference") can lead, on large scales, to different determinations of the total energy of the system. In order to accommodate this effect, we have to generalize the concept of energy, and, moreover, if we are to define the total energy contained in any universe, we must consider how to add up the energy in universes that may be infinite in spatial extent.

There is a lot of debate over precisely how to do this. The scientific literature is replete with claims and counterclaims in this regard.

One thing is certain, however: There is one universe in which the total energy is definitely and precisely zero. It is not, however, a flat universe, which is in principle infinite in spatial extent, and therefore the calculation of total energy becomes problematic. It is a closed universe, one in which the density of matter and energy is sufficient to cause space to close back upon itself. As I have described, in a closed universe, if you look far enough in one direction, you will eventually see the back of your head!

The reason the energy of a closed universe is zero is really relatively simple. It is easiest to consider the result by analogy with the fact that in a closed universe the total electric charge must also be zero.

Since the time of Michael Faraday we think of electric charge as being the source of an electric field (due in modern quantum parlance to the emission of the virtual photons I described above). Pictorially, we imagine "field lines" emanating out radi-

ally from the charge, with the number of field lines being proportional to the charge, and the direction of field lines being
outward for positive charges and inward for negative charges,
as shown below.

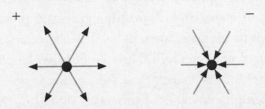

We imagine these field lines going out to infinity, and as they
spread out, getting farther apart. This implies that the strength
of the electric field gets weaker and weaker. However, in a closed
universe, the field lines associated with a positive charge, for
example, may start out spreading apart but eventually, just as
the lines of longitude on a map of the Earth come together at the
North and South Poles, the field lines from the positive charge
will come together again on the far side of the universe. When
they converge, the field will get stronger and stronger again until
there is enough energy to create a negative charge that can "eat"
the field lines at this antipodal point of the universe.

It turns out a very similar argument, in this case associated not
with the "flux" of field lines but with the "flux" of energy in a
closed universe, tells us that the total positive energy, including
that associated with the rest masses of particles, must be exactly
compensated for by a negative gravitational energy, so that the
total energy is precisely zero.

So if the total energy of a closed universe is zero, and if the
sum-over-paths formalism of quantum gravity is appropriate,
then quantum mechanically such universes could appear spontaneously with impunity, carrying no net energy. I want to empha-

size that these universes would be completely self-contained space-times, disconnected from our own.

There is a hitch, however. A closed expanding universe filled with matter will in general expand to a maximum size and then recollapse just as quickly, ending up in a space-time singularity where the no-man's land of quantum gravity at present cannot tell us what its ultimate fate will be. The characteristic lifetime of tiny closed universes will therefore be microscopic, perhaps on the order of the "Planck time," the characteristic scale over which quantum gravitational processes should operate, about 10^{-44} seconds or so.

There is a way out of this dilemma, however. If, before such a universe can collapse, the configuration of fields within it produces a period of inflation, then even an initially tiny closed universe can rapidly, exponentially expand, becoming closer and closer to an infinitely large flat universe during this period. After one hundred or so doubling times of such inflation, the universe will be so close to flat that it could easily last much longer than our universe has been around without collapsing.

Another possibility actually exists, one that always gives me a slight twinge of nostalgia (and envy), because it represented an important learning experience for me. When I was first a postdoc at Harvard, I was playing with the possible quantum mechanics of gravitational fields, and I learned of a result by a good friend from graduate school, Ian Affleck. A Canadian who had been a graduate student at Harvard when I was at MIT, Affleck joined the Society of Fellows a few years before I did and had used the mathematical theory of Feynman that we now use for dealing with elementary particles and fields, called quantum field theory, to calculate how particles and antiparticles could be produced in a strong magnetic field.

I realized that the form of the solution that Ian had described, something called an "instanton," resembled very much an inflating universe, if one took over his formalism to the case of gravity. But it looked like an inflating universe that began from nothing! Before writing up this result, I wanted to address my own confusion about how to interpret what physics such a mathematical solution might correspond to. I soon learned, however, that while I was cogitating, just down the road the very creative cosmologist I mentioned earlier, Alex Vilenkin, who has since become a friend, had actually just written a paper that described in exactly this fashion how quantum gravity indeed might create an inflating universe directly from nothing. I was scooped, but I couldn't be that upset because (a) I frankly didn't understand in detail at that point what I was doing, and (b) Alex had the boldness to propose something that at the time I didn't. I have since learned that one doesn't have to understand all the implications of one's work in order to publish. Indeed, there are several of my own most important papers that I only fully understood well after the fact.

In any case, while Stephen Hawking and his collaborator Jim Hartle have proposed a very different scheme for trying to determine the "boundary conditions" on universes that may begin from nothing at all, the important facts are these:

1. In quantum gravity, universes can, and indeed always will, spontaneously appear from nothing. Such universes need not be empty, but can have matter and radiation in them, as long as the total energy, including the negative energy associated with gravity, is zero.

2. In order for the closed universes that might be created through such mechanisms to last for longer than infinitesimal times, something like inflation is necessary. As a result, the only long-lived universe one might expect to live in as a

result of such a scenario is one that today appears flat, just as the universe in which we live appears.

The lesson is clear: quantum gravity not only appears to allow universes to be created from nothing—meaning, in this case, I emphasize, the absence of space and time—it may require them. "Nothing"—in this case no space, no time, no anything!—*is* unstable.

Moreover, the general characteristics of such a universe, if it lasts a long time, would be expected to be those we observe in our universe today.

Does this prove that our universe arose from nothing? Of course not. But it does take us one rather large step closer to the plausibility of such a scenario. And it removes one more of the objections that might have been leveled against the argument of creation from nothing as described in the previous chapter.

There, "nothing" meant empty but preexisting space combined with fixed and well-known laws of physics. Now the requirement of space has been removed.

But, remarkably, as we shall next discuss, even the laws of physics may not be necessary or required.

CHAPTER 11

BRAVE NEW WORLDS

It was the best of times. It was the worst of times.
—CHARLES DICKENS

The central problem with the notion of creation is that it appears to require some externality, something outside of the system itself, to *preexist,* in order to create the conditions necessary for the system to come into being. This is usually where the notion of God—some external agency existing separate from space, time, and indeed from physical reality itself—comes in, because the buck seems to be required to stop somewhere. But in this sense *God* seems to me to be a rather facile semantic solution to the deep question of creation. I think this is best explained within the context of a slightly different example: the origin of morality, which I first learned from my friend Steven Pinker.

Is morality external and absolute, or is it derived solely within the context of our biology and our environment, and thus can it be determined by science? During a debate on this subject organized at Arizona State University, Pinker pointed out the following conundrum.

If one argues, as many deeply religious individuals do, that without God there can be no ultimate right and wrong—namely that God determines for us what is right and wrong—one can

171

then ask the questions: What if God decreed that rape and murder were morally acceptable? Would that make them so?

While some might answer yes, I think most believers would say no, God would not make such a decree. But why not? Presumably because God would have some reason for not making such a decree. Again, presumably this is because reason suggests that rape and murder are not morally acceptable. But if God would have to appeal to reason, then why not eliminate the middleman entirely?

We may wish to apply similar reasoning to the creation of our universe. All of the examples I have provided thus far indeed involve creation of something from what one should be tempted to consider as nothing, but the *rules* for that creation, i.e., the laws of physics, were preordained. Where do the rules come from?

There are two possibilities. Either God, or some divine being who is not bound by the rules, who lives outside of them, determines them—either by whim or with malice aforethought—or they arise by some less supernatural mechanism.

The problem with God determining the rules is that you can at least ask what, or who, determined God's rules. Traditionally the response to this is to say that God is, among the Creator's many other spectacular attributes, the *cause of all causes*, in the language of the Roman Catholic Church, or the *First Cause* (as per Aquinas), or in the language of Aristotle, moving the *prime mover*.

Interestingly, Aristotle recognized the problem of a first cause, and decided that for this reason the universe must be eternal. Moreover, God himself, whom he identified as pure self-absorbed thought, the love of which motivated the prime mover to move, had to be eternal, not causing motion by creating it, but rather by establishing the end purpose of motion, which itself Aristotle deemed had to be eternal.

Aristotle felt that equating First Cause with God was less

than satisfying, in fact that the Platonic notion of First Cause was flawed, specifically because Aristotle felt every cause must have a precursor—hence, the requirement that the universe be eternal. Alternatively, if one takes the view of God as the cause of all causes, and therefore eternal even if our universe is not, the *reductio ad absurdum* sequence of "why" questions does indeed terminate, but as I have stressed, only at the expense of introducing a remarkable all-powerful entity for which there is simply no other evidence.

In this regard, there is another important point to stress here. The apparent logical necessity of First Cause is a real issue for any universe that has a beginning. Therefore, on the basis of logic alone one cannot rule out such a deistic view of nature. But even in this case it is vital to realize that this deity bears no logical connection to the personal deities of the world's great religions, in spite of the fact that it is often used to justify them. A deist who is compelled to search for some overarching intelligence to establish order in nature will not, in general, be driven to the personal God of the scriptures by the same logic.

These issues have been debated and discussed for millennia, by brilliant and not-so-brilliant minds, many of the latter making their current living by debating them. We can return to these issues now because we are simply better informed by our knowledge of the nature of physical reality. Neither Aristotle nor Aquinas knew about the existence of our galaxy, much less the Big Bang or quantum mechanics. Hence the issues they and later medieval philosophers grappled with must be interpreted and understood in the light of new knowledge.

Consider, in the light of our modern picture of cosmology, for example, Aristotle's suggestion that there are no First Causes, or rather that causes indeed go backward (and forward) infinitely far in all directions. There is no beginning, no creation, no end.

When I have thus far described how something almost always can come from "nothing," I have focused on either the creation of something from preexisting empty space or the creation of empty space from no space at all. Both initial conditions work for me when I think of the "absence of being" and therefore are possible candidates for nothingness. I have not addressed directly, however, the issues of what might have existed, if anything, before such creation, what laws governed the creation, or, put more generally, I have not discussed what some may view as the question of First Cause. A simple answer is of course that either empty space or the more fundamental nothingness from which empty space may have arisen, preexisted, and is eternal. However, to be fair, this does beg the possible question, which might of course not be answerable, of what, if anything, fixed the rules that governed such creation.

One thing is certain, however. The metaphysical "rule," which is held as an ironclad conviction by those with whom I have debated the issue of creation, namely that "*out of nothing nothing comes*," has no foundation in science. Arguing that it is self-evident, unwavering, and unassailable is like arguing, as Darwin falsely did, when he made the suggestion that the origin of life was beyond the domain of science by building an analogy with the incorrect claim that matter cannot be created or destroyed. All it represents is an unwillingness to recognize the simple fact that nature may be cleverer than philosophers or theologians.

Moreover, those who argue that out of nothing nothing comes seem perfectly content with the quixotic notion that somehow God can get around this. But once again, if one requires that the notion of true nothingness requires not even the *potential* for existence, then surely God cannot work his wonders, because if he does cause existence from nonexistence, there must have been the potential for existence. To simply argue that God can do what nature cannot is to argue that *supernatural* potential for existence is somehow different from regular natural potential for existence. But this seems an arbitrary semantic distinction designed by those who have decided in advance (as theologians are wont to do) that the supernatural (i.e., God) must exist so they define their philosophical ideas (once again completely divorced from any empirical basis) to exclude anything but the possibility of a god.

In any case, to posit a god who could resolve this conundrum, as I have emphasized numerous times thus far, often is claimed to require that God exists outside the universe and is either timeless or eternal.

Our modern understanding of the universe provides another plausible and, I would argue, far more physical solution to this problem, however, which has some of the same features of an external creator—and moreover is logically more consistent.

I refer here to the multiverse. The possibility that our universe is one of a large, even possibly infinite set of distinct and causally separated universes, in each of which any number of fundamental aspects of physical reality may be different, opens up a vast new possibility for understanding our existence.

As I have mentioned, one of the more distasteful but potentially true implications of these pictures is that physics, at some fundamental level, is merely an environmental science. (I find this distasteful because I was brought up on the idea that the goal of science was to explain why the universe had to be the way it is and how that came to be. If instead the laws of physics as we know them are merely accidents correlated to our existence, then that fundamental goal was misplaced. However, I will get over my prejudice if the idea turns out to be true.) In this case, the fundamental forces and constants of nature in this picture are no more fundamental than the Earth-Sun distance. We find ourselves living on Earth rather than Mars not because there is something profound and fundamental about the Earth-Sun distance, but rather simply if Earth were located at a different distance, then life as we know it could not have evolved on our planet.

These anthropic arguments are notoriously slippery, and it is almost impossible to make specific predictions based on them without knowing explicitly both the probability distribution among all possible universes of the various fundamental constants and forces—namely, which may vary and which don't, and what possible values and forms they may take—and also exactly how "typical" we are in our universe. If we are not "typical" life forms, then anthropic selection, if it occurs at all, may be based on different factors from those we would otherwise attribute it to.

Nevertheless, a multiverse, either in the form of a landscape of universes existing in a host of extra dimensions, or in the form of a possibly infinitely replicating set of universes in a three-dimensional space as in the case of eternal inflation, changes the playing field when we think about the creation of our own universe and the conditions that may be required for that to happen.

In the first place, the question of what determined the laws of nature that allowed our universe to form and evolve now becomes less significant. If the laws of nature are themselves stochastic and random, then there is no prescribed "cause" for our universe. Under the general principle that anything that is not forbidden is allowed, then we would be guaranteed, in such a picture, that some universe would arise with the laws that we have discovered. No mechanism and no entity is required to fix the laws of nature to be what they are. They could be almost anything. Since we don't currently have a fundamental theory that explains the detailed character of the landscape of a multiverse, we cannot say. (Although to be fair, to make any scientific progress in calculating possibilities, we generally assume that certain properties, like quantum mechanics, permeate all possibilities. I have no idea if this notion can be usefully dispensed with, or at least I don't know of any productive work in this regard.)

In fact, there may be no fundamental theory at all. Although I became a physicist because I hoped that there was such a theory, and because I hoped that I might one day help contribute to discovering it, this hope may be misplaced, as I have already lamented. I take solace in the statement by Richard Feynman, which I summarized briefly before, but want to present in its entirety here:

> People say to me, "Are you looking for the ultimate laws of physics?" No, I'm not. I'm just looking to find out more about the world, and if it turns out there is a simple ultimate law that explains everything, so be it. That would be very nice to discover. If it turns out it's like an onion with millions of layers, and we're sick and tired of looking at layers, then that's the way it is . . . My interest in science is to simply find out more about the world, and the more I find out, the better it is. I like to find out.

One can carry the argument further and in a different direction, which also has implications for the arguments at the core of this book. In a multiverse of any of the types that have been discussed, there could be an infinite number of regions, potentially infinitely big or infinitesimally small, in which there is simply "nothing," and there could be regions where there is "something." In this case, the response to why there is something rather than nothing becomes almost trite: there is something simply because if there were nothing, we wouldn't find ourselves living there!

I recognize the frustration inherent in such a trivial response to what has seemed such a profound question throughout the ages. But science has told us that anything profound or trivial can be dramatically different from what we might suppose at first glance.

The universe is far stranger and far richer—more wondrously strange—than our meager human imaginations can anticipate. Modern cosmology has driven us to consider ideas that could not even have been formulated a century ago. The great discoveries of the twentieth and twenty-first centuries have not only changed the world in which we operate, they have revolutionized our understanding of the world—or worlds—that exist, or may exist, just under our noses: the reality that lies hidden until we are brave enough to search for it.

This is why philosophy and theology are ultimately incapable of addressing by themselves the truly fundamental questions that perplex us about our existence. Until we open our eyes and let nature call the shots, we are bound to wallow in myopia.

Why is there something rather than nothing? Ultimately, this question may be no more significant or profound than asking why some flowers are red and some are blue. "Something" may always come from nothing. It may be required, independent of the underlying nature of reality. Or perhaps "something" may not be very special or even very common in the multiverse. Either way, what is really useful is not pondering this question, but rather participating in the exciting voyage of discovery that may reveal specifically how the universe in which we live evolved and is evolving and the processes that ultimately operationally govern our existence. That is why we have science. We may supplement this understanding with reflection and call that philosophy. But only via continuing to probe every nook and cranny of the universe that is accessible to us will we truly build a useful appreciation of our own place in the cosmos.

Before concluding, I want to raise one more aspect of this question that I haven't touched upon, but which strikes me as worth ending with. Implicit in the question of why there is something rather than nothing is the solipsistic expectation that "something" will persist—that somehow the universe has "progressed" to the point of our existence, as if we were the pinnacle of creation. Far more likely, based on everything we know about the universe, is the possibility that the future, perhaps the infinite future, is one in which nothingness will once again reign.

If we live in a universe whose energy is dominated by the energy of nothing, as I have described, the future is indeed bleak. The heavens will become cold and dark and empty. But the situation is actually worse. A universe dominated by the energy of empty space is the worst of all universes for the future of life. Any civilization is guaranteed to ultimately disappear in such a universe, starved of energy to survive. After an unfathomably long time, some quantum fluctuation or some thermal agitation may produce a local region where once again life can evolve and thrive. But that too will be ephemeral. The future will be dominated by a universe with nothing in it to appreciate its vast mystery.

Alternatively, if the matter that makes us up was created at the beginning of time by some quantum processes, as I have described, we are virtually guaranteed that it, too, will disappear once again. Physics is a two-way street, and beginnings and endings are linked. Far, far into the future, protons and neutrons will decay, matter will disappear, and the universe will approach a state of maximum simplicity and symmetry.

Mathematically beautiful perhaps, but devoid of substance.

As Heraclitus of Ephesus wrote in a slightly different context, "Homer was wrong in saying: 'Would that strife might perish from among gods and men!' He did not see that he was praying for the destruction of the universe; for if his prayers were heard, all things would pass away." Or, as Christopher Hitchens has restated it, "Nirvana *is* nothingness."

A more extreme version of this eventual retreat into nothingness may be inevitable. Some string theorists have argued, on the basis of complex mathematics, that a universe like ours, with a positive energy in empty space, *cannot* be stable. Eventually, it must decay to a state in which the energy associated with space will be negative. Our universe will then recollapse inward to a point, returning to the quantum haze from which our own existence may have begun. If these arguments are correct, our universe will then disappear as abruptly as it probably began.

In this case, the answer to the question, "Why is there something rather than nothing?" will then simply be: "There won't be for long."

Epilogue

The sanction of experienced fact as a face of truth is a profound subject, and the mainspring which has moved our civilization since the Renaissance.

—Jacob Bronowski

I began this book with another quote from Jacob Bronowski:

Dream or nightmare, we have to live our experience as it is, and we have to live it awake. We live in a world which is penetrated through and through by science and which is both whole and real. We cannot turn it into a game simply by taking sides.

As I have also argued, one person's dream is another person's nightmare. A universe without purpose or guidance may seem, for some, to make life itself meaningless. For others, including me, such a universe is invigorating. It makes the fact of our existence even more amazing, and it motivates us to draw meaning from our own actions and to make the most of our brief existence in the sun, simply because we are here, blessed with consciousness and with the opportunity to do so. Bronowski's point, however, is that it doesn't really matter either way, and what we would like for the universe is irrelevant. Whatever happened, happened, and it happened on a cosmic scale. And whatever is about to happen on that scale will happen independent of our likes and dislikes. We cannot affect the former, and we are unlikely to affect the latter.

What we can do, however, is try to understand the circumstances of our existence. I have described in this book one of the most remarkable journeys of exploration humanity has ever taken in its evolutionary history. It is an epic quest to explore and understand the cosmos on scales that simply were unknown a century ago. The journey has pushed the limits of the human spirit, combining the willingness to follow evidence wherever it might lead with the courage to devote a lifetime to exploring the unknown with the full knowledge that the effort might go nowhere, and finally requiring a mixture of creativity and persistence to address the often tedious tasks of sorting through endless equations or endless experimental challenges.

I have always been attracted to the myth of Sisyphus and have likened the scientific effort at times to his eternal task of pushing a boulder up a mountain, only to have it fall back each time before he reaches the top. As Camus imagined, Sisyphus was smiling, and so should we. Our journey, whatever the outcome, provides its own reward.

The phenomenal progress we have made in the past century has brought us to the cusp, as scientists, of operationally addressing the deepest questions that have existed since we humans took our first tentative steps to understand who we are and where we came from.

As I have described here, in the process the very meaning of these questions has evolved along with our understanding of the universe. "Why is there something rather than nothing?" must be understood in the context of a cosmos where the meaning of these words is not what it once was, and the very distinction between something and nothing has begun to disappear, where transitions between the two in different contexts are not only common, but required.

As such, the question itself has been sidelined as we strive in our quest for knowledge. Instead, we are driven to understand the processes that govern nature in a way that allows us to make predictions and, whenever possible, to affect our own future. In so doing, we have discovered that we live in a universe in which empty space—what formerly could have passed for nothing— has a new dynamic that dominates the current evolution of the cosmos. We have discovered that all signs suggest a universe that could and plausibly did arise from a deeper nothing—involving the absence of space itself—and which may one day return to nothing via processes that may not only be comprehensible but also processes that do not require any external control or direction. In this sense, science, as physicist Steven Weinberg has emphasized, does not make it impossible to believe in God, but rather makes it possible to not believe in God. Without science, everything is a miracle. With science, there remains the possibility that nothing is. Religious belief in this case becomes less and less necessary, and also less and less relevant.

The choice to turn to the notion of divine creation falls to each of us, of course, and I don't expect the ongoing debate to die down anytime soon. But as I have stressed, I believe that if we are to be intellectually honest, we must make an informed choice, informed by fact, not by revelation.

That has been the purpose of this book, to provide an informed picture of the universe as we understand it and to describe the theoretical speculations that currently are driving physics forward as we scientists attempt to separate the wheat from the chaff in our observations and theories.

I have made clear my own predilection: the case that *our* universe arose from nothing seems by far the most compelling intellectual alternative to me at the present time. You will draw your own conclusion.

I want to end my discussion by returning to a question that I personally find even more intellectually fascinating than the question of something from nothing. It is the question Einstein asked about whether God had any choice in the creation of the universe. This question provides the basic motivation for almost all research into the fundamental structure of matter, space, and time—the research that has occupied me for much of my professional life.

I used to think there was a stark choice in the answer to this question, but in the process of writing this book, my views have altered. Clearly, if there is a single theory involving a unique set of laws that describes and, indeed, prescribes how our universe came into being and the rules that have governed its evolution ever since—the goal of physics since Newton or Galileo—then the answer would appear to be, "No, things had to be the way they were, and are."

But if our universe is not unique, and it is a part of a vast and possibly infinite multiverse of universes, would the answer to Einstein's question be a resounding "Yes, there is a host of choices for existence"?

I am not so sure. It could be that there is an infinite set of different combinations of laws and varieties of particles and substances and forces and even distinct universes that may arise in such a multiverse. It may be that only a certain very restricted combination, one that results in the universe of the type in which we live or one very much like it, can support the evolution of beings who can ask such a question. Then the answer to Einstein will still remain negative. A God or a Nature that could encompass a multiverse would be as constrained in the creation of a universe in which Einstein could ask the question as either would be if there is only one choice of a consistent physical reality.

I find oddly satisfying the possibility that, in either scenario, even a seemingly omnipotent God would have no freedom in the creation of our universe. No doubt because it further suggests that God is unnecessary—or at best redundant.

Afterword

by Richard Dawkins

Nothing expands the mind like the expanding universe. The music of the spheres is a nursery rhyme, a jingle to set against the majestic chords of the Symphonie Galactica. Changing the metaphor and the dimension, the dusts of centuries, the mists of what we presume to call "ancient" history, are soon blown off by the steady, eroding winds of geological ages. Even the age of the universe, accurate—so Lawrence Krauss assures us—to the fourth significant figure at 13.72 billion years, is dwarfed by the trillennia that are to come.

But Krauss's vision of the cosmology of the remote future is paradoxical and frightening. Scientific progress is likely to go into reverse. We naturally think that, if there are cosmologists in the year 2 trillion AD, their vision of the universe will be expanded over ours. Not so—and this is one of the many shattering conclusions I take away on closing this book. Give or take a few billion years, ours is a very propitious time to be a cosmologist. Two trillion years hence, the universe will have expanded so far that all galaxies but the cosmologist's own (whichever one it happens to be) will have receded behind an Einsteinian horizon so absolute, so inviolable, that they are not only invisible but beyond all possibility of leaving a trace, however indirect.

They might as well never have existed. Every trace of the Big Bang will most likely have gone, forever and beyond recovery. The cosmologists of the future will be cut off from their past, and from their situation, in a way that we are not.

We know we are situated in the midst of 100 billion galaxies, and we know about the Big Bang because the evidence is all around us: the redshifted radiation from distant galaxies tells us of the Hubble expansion and we extrapolate it backward. We are privileged to see the evidence because we look out on an infant universe, basking in that dawn age when light can still travel from galaxy to galaxy. As Krauss and a colleague wittily put it, "We live at a very special time . . . the only time when we can observationally verify that we live at a very special time!" The cosmologists of the third trillennium will be forced back to the stunted vision of our early twentieth century, locked as we were in a single galaxy which, for all that we knew or could imagine, was synonymous with the universe.

Finally, and inevitably, the flat universe will further flatten into a nothingness that mirrors its beginning. Not only will there be no cosmologists to look out on the universe, there will be nothing for them to see even if they could. Nothing at all. Not even atoms. Nothing.

If you think that's bleak and cheerless, too bad. Reality doesn't owe us comfort. When Margaret Fuller remarked, with what I imagine to have been a sigh of satisfaction, "I accept the universe," Thomas Carlyle's reply was withering: "Gad, she'd better!" Personally, I think the eternal quietus of an infinitely flat nothingness has a grandeur that is, to say the least, worth facing off with courage.

But if something can flatten into nothing, can nothing spring into action and give birth to something? Or why, to quote a theological chestnut, is there something rather than nothing?

Here we come to perhaps the most remarkable lesson that we are left with on closing Lawrence Krauss's book. Not only does physics tell us how something could have come from nothing, it goes further, by Krauss's account, and shows us that nothingness is unstable: something was almost bound to spring into existence from it. If I understand Krauss aright, it happens all the time: The principle sounds like a sort of physicist's version of two wrongs making a right. Particles and antiparticles wink in and out of existence like subatomic fireflies, annihilating each other, and then re-creating themselves by the reverse process, out of nothingness.

The spontaneous genesis of something out of nothing happened in a big way at the beginning of space and time, in the singularity known as the Big Bang followed by the inflationary period, when the universe, and everything in it, took a fraction of a second to grow through twenty-eight orders of magnitude (that's a 1 with twenty-eight zeroes after it—think about it).

What a bizarre, ridiculous notion! Really, these scientists! They're as bad as medieval Schoolmen counting angels on pinheads or debating the "mystery" of the transubstantiation.

No, not so, not so with a vengeance and in spades. There is much that science still doesn't know (and it is working on it with rolled-up sleeves). But some of what we do know, we know not just approximately (the universe is not mere thousands but billions of years old): we know it with confidence and with stupefying accuracy. I've already mentioned that the age of the universe is measured to four significant figures. That's impressive enough, but it is nothing compared to the accuracy of some of the predictions with which Lawrence Krauss and his colleagues can amaze us. Krauss's hero Richard Feynman pointed out that some of the predictions of quantum theory—again based on assumptions that seem more bizarre than anything dreamed up by even the most obscurantist of theologians—have been verified with such accuracy that they are equivalent to predicting the distance between New York and Los Angeles to within one hairsbreadth.

Theologians may speculate about angels on pinheads or whatever is the current equivalent. Physicists might seem to have their own angels and their own pinheads: quanta and quarks, "charm," "strangeness," and "spin." But physicists can count their angels and can get it right to the nearest angel in a total of 10 billion: not an angel more, not an angel less. Science may be weird and incomprehensible—more weird and less comprehensible than any theology—but science works. It gets results. It can fly you to Saturn, slingshotting you around Venus and Jupiter on the way. We may not understand quantum theory (heaven knows, I don't), but a theory that predicts the real world to ten decimal places cannot in any straightforward sense be wrong. Theology not only lacks decimal places: it lacks even the smallest hint of a connection with the real world. As Thomas Jefferson said, when founding his University of Virginia, "A professorship of Theology should have no place in our institution."

If you ask religious believers why they believe, you may find a few "sophisticated" theologians who will talk about God as the "Ground of all Isness," or as "a metaphor for interpersonal fellowship" or some such evasion. But the majority of believers leap, more honestly and vulnerably, to a version of the argument from design or the argument from first cause. Philosophers of the caliber of David Hume didn't need to rise from their armchairs to demonstrate the fatal weakness of all such arguments: they beg the question of the Creator's origin. But it took Charles Darwin, out in the real world on HMS *Beagle,* to discover the brilliantly simple—and non-question-begging—alternative to design. In the field of biology, that is. Biology was always the favorite hunting ground for natural theologians until Darwin—

not deliberately, for he was the kindest and gentlest of men—chased them off. They fled to the rarefied pastures of physics and the origins of the universe, only to find Lawrence Krauss and his predecessors waiting for them.

Do the laws and constants of physics look like a finely tuned put-up job, designed to bring us into existence? Do you think some agent must have caused everything to start? Read Victor Stenger if you can't see what's wrong with arguments like that. Read Steven Weinberg, Peter Atkins, Martin Rees, Stephen Hawking. And now we can read Lawrence Krauss for what looks to me like the knockout blow. Even the last remaining trump card of the theologian, "Why is there something rather than nothing?" shrivels up before your eyes as you read these pages. If *On the Origin of Species* was biology's deadliest blow to supernaturalism, we may come to see *A Universe from Nothing* as the equivalent from cosmology. The title means exactly what it says. And what it says is devastating.

INDEX

Index

Index

as possibly infinite, xiii
quantum theory as applying to, 161
space-time, quantum mechanics, 162, 163
spectrum, 9–10, 59, 66
Spinoza, Baruch, 160
spiral galaxies (spiral nebulae), 7, 17
absorption lines of, 10–11
"standard candle," 19, 78–79
stars:
brightness of, 7–8, 79
composition of, 10, 20
elements made in, 17–19, 113–14
main sequence, 107
variable, 7
Star Trek, 61
static electricity, 154
Stenger, Victor, 191
string theory, 2, 130–35, 180
sum over paths formalism, 162–63, 167–69
Sun, spectrum of, 9
superclusters, 28
superconductivity, 132
supernaturalism, 141–42
supernova:
Brahe's observation of, 20
elements created in, 17–18
number of, 17, 20–21
and rate of expansion of universe, 80–82, 84–86, 88
Type 1a, 19, 79, 80–82, 84–86, 88
Supernova Cosmology Project, 81–82
superstrings, 132
supersymmetry, 132
symmetry, 73, 132, 179
Sysiphus, 182

tauons, 136
theology, xi, xiii—xiv, xvi, 144, 145–46, 178, 190
see also God
Theory of Everything, 132, 134
"Theory of Positrons, A" (Feynman), 64–65

Thomas Aquinas, xii, 135, 149, 172, 173
thought experiments, 2
time:
antiparticles as appearing to move backward in, 62–65
as arising from nothing, xiv—xv
characteristic scale imprinted on last scattering surface by, 44–45
in general theory of relativity, 1, 161
as possibly infinite, xiii
see also space-time
total energy:
in closed universe, 166–68
in flat universe, 165
problems in measurement of, 166
as total gravitational energy plus energy associated with mass, 165
total gravitational energy, 101–3, 105
definition of, 100
in inflation, 101–3
Newtonian equation for, 103
as nonaribtrary, 148–49
in total energy, 165
in universe arising from nothing, 149–52, 165
triangles, sum of angles in, 39–40, 50
Turner, Michael, 56, 75–76, 78, 81
Type 1a supernova, 19, 79, 80–82, 84–86, 88
Tyson, Tony, 32–34

Uncertainty Principle, 59, 62–65, 71, 156, 164, 165
universe:
age of, 3, 15–16, 21, 42, 77, 86–87, 92, 187, 189
alleged rotation of, 84
average density of, 123
boundary conditions of creation from nothing of, 169–70
cooling of, 95, 96
end of, 23, 26, 28
as eternal, 172–74
homogeneity of, 97, 115

201

ABOUT THE AUTHOR

Lawrence M. Krauss is Foundation Professor in the School of Earth and Space Exploration and the Physics Department at Arizona State University, as well as Co-Director of the Cosmology Initiative and Inaugural Director of the Origins Project. The Origins program involves new and wide-ranging interdisciplinary research, teaching, and outreach focusing on all aspects of origins: from the origins of the cosmos to human origins, to the origins of consciousness and culture. Krauss is an internationally known theoretical physicist with broad research interests, including the interface between elementary particle physics and cosmology. He received his PhD in physics from the Massachusetts Institute of Technology in 1982, and joined the Harvard Society of Fellows. In 1985, he joined the faculty of physics at Yale University, and then moved to Case Western Reserve University as Ambrose Swasey Professor in 1993. From 1993 to 2005, he served as chairman of the physics department at Case. He is the recipient of numerous international awards for his research and writing, and is the only physicist to receive awards from all three major US physics societies, the American Physical Society, the American Association of Physics Teachers, and the American Institute of Physics.

Krauss is also one of the few prominent scientists today to have actively crossed the chasm between science and popular culture, causing him to be heralded as a unique "public intellectual" by *Scientific American* magazine. For example, besides his books and radio and television work, and his newspaper and magazine commentaries, Krauss has performed solo with the Cleveland Orchestra, narrating Gustav Holst's *The Planets* at the Blossom Music Center in the most highly attended concert at that venue, and he was nominated for a Grammy Award for his liner notes for a Telarc CD of music from *Star Trek*. In 2005, he also served as a jury member at the Sundance Film Festival.

Q & A with the Author

1. **What do you really mean by "nothing"?**

    As I describe in the book, I believe it is most useful to base
    our definitions on empirically discovered realities rather
    than abstract philosophical precepts. For me, independent
    of the question of "non-existence," which takes one off
    on lots of potentially deep philosophical issues but rather
    impotent physics ideas, the really seemingly miraculous
    aspect of our universe, which I also believe has inspired
    much of the debate about this topic over the centuries, is
    how all of the stuff we see could have arisen from a universe
    in which that stuff did not already exist. It seems at the
    very least to violate the conservation of energy, and more
    significantly, common sense. But one of the great things
    about science that I want to convey is that common sense
    is not necessarily a good guide to understanding nature at
    the forefront. Our common sense should derive from the
    universe, rather than vice versa. And the remarkable non-
    miraculous miracle is that combining quantum mechanics
    with gravity allows stuff to arise from no-stuff.

    Now, that state of no-stuff may not be "nothing" in a
    classical sense, but it is a remarkable transformation never-
    theless. So, the first form of "nothing" is just empty space.
    But one is perfectly reasonable in questioning whether

this is really "nothing" because space is there, as is time. I then describe how it is possible that space and time themselves could have arisen from no space and time, which is certainly closer to absolute nothing. Needless to say, one can nevertheless question whether that is nothing, because the transition is mediated by some physical laws. Where did they come from? That is a good question, and one of the more modern answers is that even the laws themselves may be random, coming into existence along with universes that may arise. This may still beg the question of what allows any of this to be possible, but at some level it is, as I describe at the beginning of the book, "turtles all the way down." There are questions we can address effectively via empirical methods and questions we can ask that don't lead to physical insights and predictions. The trick is to tell the difference between the two.

2. **Why "How?" and not "Why?"**

"Why" questions are laced with intellectual baggage that is usually unintended. We can ask "Why are there nine planets around our sun?" (since for me Pluto will always be a planet!) but by that we don't ascribe significance or purpose to the number nine, as if the universe was designed so there would be nine planets around the sun. If our sun was the only star, then one might ascribe some significance to that particular number (as Kepler did when he tried to explain six planets in terms of Platonic solids). But we are much more interested in the different ways solar systems can arise and how. Asking "Why?" presumes some purpose, which need not exist. Ultimately one can keep asking why forever, and the ultimate answer may simply be "Because," but that doesn't illuminate much.

3. **If you get rid of God, then does life lose all purpose?**

For me it certainly doesn't—quite the opposite. I would find little purpose living in a world ruled by some divine Saddam Hussein—like character, as my late friend Christopher Hitchens put it, who not only makes all the rules, but punishes those who disobey them with eternal damnation. I find living in a universe without purpose to be amazing, because it makes the accident of our existence and our consciousness even more precious—something to be valued during our brief moment in the sun.

4. **What do you mean by "flat"? Is the universe flat as a pancake?**

I wish I had described this a little more carefully in the hardcover edition, and I have expanded the discussion in this edition. A flat three-dimensional space is just the kind of space you already thought you lived in, where light rays travel in straight lines, and perpendicular axes (x, y, and z) remain perpendicular. In curved three-dimensional spaces neither statement is true. Since mass and energy can curve space locally (i.e., around the sun and earth for example), the big question is what about the global structure of space on the largest scales: Is it curved or not? And it turns out on the largest observable scales, it isn't. And that fact is very telling, as I describe in the book, because it is what one would expect from a universe that arose from nothing.

5. **Isn't science just another kind of faith?**

Absolutely not. Scientists change their minds, admit they are wrong, and are happy and eager to throw out ideas that turn out not to work. We don't presume to know for certain the answers to questions before we ask them. So yes, we have faith that the universe is comprehensible, but the greatest thing about science is that our faith is shakable. At any moment we can give up believing in anything we once believed in, if nature suggests otherwise.

6. **Does the search for the Higgs boson at the Large Hadron Collider have cosmological significance? What if we discover it? What if we don't?**

I discuss this issue in the new preface to this paperback edition. The search for the Higgs boson reflects the capstone of a remarkable intellectual journey that began over fifty years ago, and if it is discovered at the LHC as initial results reported in 2011 suggest, it would validate a theoretical edifice that otherwise would be on shaky ground. In that sense it would be remarkable if our ideas about the Higgs are correct, because usually nature surprises us. Most theories are in fact wrong. If that weren't the case, anyone could do physics. But in any case if the Higgs exists, it means that another aspect of our existence is a cosmic accident. Particles would get their masses by interactions with a background, otherwise invisible, field, like trying to swim through molasses. That means if such a field had not become established in the early universe, we wouldn't be here . . . yet more something from nothing! At the same time, a Higgs discovery at the LHC will likely raise more questions than answers: Why does it have the mass it does? How can we understand its existence in the context of all four known forces in nature? And so on.

7. **I have read it claimed that the fundamental laws of nature have nothing to say on the subject of where observed forces came from, or of why the world should**

have consisted of the particular kinds of particles and fields it does, or why there should have been a world in the first place. Can you comment?

In fact, it is one of the great developments in particle physics in the past forty years to realize that the properties of the universe we see, which forces are manifest, which particular kinds of fields can exist on observable scales, and which particles have mass and which don't, can arise spontaneously as an accident of our circumstances. This phenomenon is called "spontaneous symmetry breaking," and it basically says that as the universe evolves and cools, some background field can develop throughout space, just like an ice crystal spontaneously forms on your window sill and just as the Higgs field is predicted to have done. (The nature of the specific patterns on your window sill on a frosty day is not predetermined at the beginning of time but arises dynamically.)

When this background field develops, it causes some particles to become massive (and therefore become unstable to decay to other particles and disappear) and others to remain massless. It also determines which forces operate at long distances, like electromagnetism, and which don't, like the weak interaction. As for why a world can exist in the first place, once again, spontaneous symmetry breaking—in this case including the possibility of gravity acting—can cause some universes to expand indefinitely and be long lived, while others will disappear in an instant. Thus it can also explain why some worlds exist long enough to ask the question: "Why is there something rather than nothing?"

8. Isn't it presumptuous to claim that we know the universe came from nothing, and that science has answered all the outstanding questions of cosmology?

It is amusing to read this criticism, usually launched by people who haven't read the book. One of the central points of my book is that we *don't* know all the answers, but what we have learned is remarkably tantalizing, while at the same time butting us up against some profound fundamental questions that may never be truly amenable to empirical falsification.

9. **Isn't science compatible with religion? After all, they both explore the same questions, don't they?**

    Science is compatible with some basic form of deism—namely, we cannot say that a universe, even one that comes from nothing by natural physical processes, was not created with some underlying purpose that may not be evident. (The fact that there is no evidence of purpose makes it of course harder to argue for one, but never mind.) But having said that, science is not compatible with all the strict doctrines of all the world's major religions, and that includes Christianity, Judaism, Islam, as well as some of the minor ones, like Mormonism and Buddhism. And there is good reason for this: The doctrines were written down by people who didn't know how the world worked. Except for Mormonism, which is recent, they were written down when we didn't know that the Earth orbited the Sun!

10. **Are you an atheist?**

    Not in the sense that I can claim definitively that there is no God or purpose to the universe. I cannot claim definitively that there isn't a teapot orbiting Jupiter, as Bertrand Russell once said. It is highly unlikely, of course. But what I can claim definitively is that I wouldn't want to live in a universe with a God—that makes me an anti-theist, as my friend Christopher Hitchens was.